Light-Driven Materials and Devices - Fundamentals and Emerging Applications

Edited by Murat Kabatas

Published in London, United Kingdom

Light-Driven Materials and Devices - Fundamentals and Emerging Applications
http://dx.doi.org/10.5772/intechopen.1006437
Edited by Murat Kabatas

Contributors
Anran Guo, Badra Bouabdallah, Beichen Liu, Durre Sameen, Faryal Idrees, Harish Santhana Krishnan, Hassan Abdoos, Jayaram Archana, Jiyuan Zheng, Jizhe Zhao, Lai Wang, Maryam Hosseinpour, Naskath Sithara H, Navaneethan Mani, Rui Wu, Saliha Kheris, Saqib Ahmad, Vinoth Elangovan, Xiaoran Wang, Xiayang Hua, Yinjie Liu, Yubo Yang, Zakia Nabi, Zhibiao Hao

First published in London, United Kingdom, 2025 by IntechOpen
IntechOpen is the global imprint of INTECHOPEN LIMITED, registered in England and Wales, registration number: 11086078, 167-169 Great Portland Street, London, W1W 5PF, United Kingdom

For EU product safety concerns: IN TECH d.o.o., Prolaz Marije Krucifikse Kozulić 3, 51000 Rijeka, Croatia, info@intechopen.com or visit our website at intechopen.com.

British Library Cataloguing-in-Publication Data
A catalogue record for this book is available from the British Library

Light-Driven Materials and Devices - Fundamentals and Emerging Applications
Edited by Murat Kabatas
p. cm.
Print ISBN 978-1-83634-770-5
Online ISBN 978-1-83634-769-9
eBook (PDF) ISBN 978-1-83634-771-2

Meet the editor

Murat Kabatas is a researcher specializing in nanomaterial chemistry and physics, focusing on gas sensors and optoelectronic devices. His work involves synthesizing and characterizing Ohmic and Schottky heterojunction thin films using chemical and physical deposition techniques. He has contributed to developing advanced sensing materials and photodetectors, with a strong interest in semiconductor interfaces and nanoscale material engineering. His interdisciplinary approach bridges functional materials science with practical device applications.

Contents

Preface

This edited volume brings together a collection of chapters that reflect the latest developments and research trends in semiconductors, photodetectors, and gas sensing technologies. The selected works span fundamental principles, innovative materials, advanced device designs, and emerging light and gas detection applications.

Chapter 1, *Semiconductors: History and Applications*, offers a comprehensive overview of the history, development, and diverse applications of semiconductors. This sets the stage for the specialized topics covered in the following chapters.

Chapter 2, *Advances in Photovoltaic Detectors: Principles, Challenges, and the Role of Avalanche Detectors in High-Sensitivity Detection*, examines the underlying principles of photovoltaic detectors. They discuss the challenges associated with their design and highlight the important role of avalanche detectors in achieving high sensitivity.

Chapter 3, *2D Materials and their Application to Near Infrared Detection*, explores the use of two-dimensional materials in near-infrared detection. The chapter discusses the unique properties of these materials and their potential for developing next-generation photodetectors.

Chapter 4, *Understanding Photodiodes through Simulation: A Case Study of GaAs PIN Photodiodes*, presents a simulation-based approach to understanding GaAs PIN photodiodes. The authors provide a detailed case study that offers valuable insights into the behavior and performance of these devices.

Finally, *Chapter 5, Optimal Photodeposition of Co-Catalyst Over 2D Heterostructures for Improved H_2 Production/Water-Splitting*, explores the role of co-catalyst deposition on 2D heterostructures to enhance hydrogen production and water splitting efficiency. This chapter links material science advancements to practical applications in clean energy technologies.

We thank all the authors for their dedicated efforts in preparing their contributions. Their expertise and insights have been instrumental in shaping this volume. We are also grateful to the editorial team for their valuable support.

We hope this collection will serve as a useful resource for researchers, engineers, and students, providing both foundational knowledge and inspiration for further research in semiconductors, photodetectors, and gas sensing technologies.

Murat Kabatas
Precision and Microsystems Engineering Department,
Delft University of Technology,
Delft, The Netherlands

Chapter 1

Semiconductors: History and Applications

Maryam Hosseinpour and Hassan Abdoos

Abstract

The development of semiconductor physics and technology has led to significant changes in our daily life and perhaps today it is impossible to live without them. Semiconductors are now widely used in many aspects of modern human civilization, including energy, sensing, electronics, biomedical, imaging, water treatment, etc. This chapter will cover semiconductors' definition, characteristics, history, and fundamental principles, emphasizing their band structure, physics, and classification. Following that, the applications of semiconductors, particularly in the energy, electronics, biomedical, and photocatalysis processes will be discussed. Following that, heterostructured photocatalysts as a possible way to increase photocatalysis performance and introduce different types of heterostructured semiconductors will be reviewed. Then the manufacturing process and semiconductor devices are discussed. The final section of the chapter will go over the effect of nanotechnology on semiconductor properties and the future trend of semiconductors. This chapter attempts to present a brief and comprehensive understanding of the history, physics, and practical application of semiconductors, with a focus on heterogeneous compounds for photocatalytic applications.

Keywords: semiconductors, intrinsic and extrinsic, band gap engineering, electron-hole pairs, photocatalysis, heterostructured nanocomposites

1. Introduction

1.1 Definition, characteristics

The term semiconductor is composed of two words: semi and conductor. Semi means imperfect while conductive refers to something that can conduct electricity. In this regard, semiconductors are materials with properties between conductors and insulators that have the capability of conducting electric current.

The temperature dependency of semiconductor conductivity is one of their most crucial characteristics. Metals and semiconductors are distinguished by this feature. While the conductivity of metals decreases with increasing temperature, the conductivity of semiconductors increases with increasing temperature. Another outstanding characteristic of semiconductors is the possibility of adjusting their energy band gap (Eg), because the range of semiconductor applications is determined by this parameter.

The interatomic bonding configurations, different types of defects or imperfections, and structural properties are among the factors that are necessary for understanding and describing semiconductors behaviors. For example, amorphous, crystalline, and polycrystalline semiconductors have found their applications in different areas including energy or electronic applications. Another unique feature of semiconductors is the ability to manipulate their structure through the doping process. By altering the structure, conductivity, and energy gap, doping has made it possible to manufacture a wide range of electronic components and numerous uses in daily life.

1.2 History

Semiconductors have a long and complex history. Semiconductor research began inconspicuously around 120 years ago. Although the word "semiconductor" was introduced by Königsberg and Weiss in 1911, it was previously used similarly by Ebert in 1789 and Bromme in 1851 [1]. However, Alessandro Volta originated the word "semiconducting" in 1782 [2] and perhaps the first notable observation in the semiconductor field was made by Michael Faraday in 1833 when he discovered that silver sulfide exhibited a negative temperature coefficient of resistance [3]. Smith discovered selenium's photoconductivity in 1973. Braun found in 1874 that contacts between some metal sulfides and metal tips showed varying electrical resistance when the applied voltage's polarity was reversed. Although Braun's finding was not immediately recognized, it ultimately contributed significantly to the advancement of radio technology and the detection of microwave radiation in World War II radar systems [4].

In 1879, the Hall Effect was discovered, although the electron had not yet been identified at that time. The word "semiconductor" was introduced in 1911. Grondahl first employed copper oxide as a rectifier in 1926, followed by selenium rectifiers. In 1931, A. H. Wilson [5] presented a quantum mechanical model of a solid semiconductor, which has since been essential for understanding semiconductor's behavior. In 1931, H. Dember discovered a potential difference between illuminated and unilluminated regions of a cuprous oxide wafer. Lang developed the first practical application of copper oxide in photocells in 1932. In 1939, Si-based microwaves were produced, followed by the discovery of IR detectors of sulfide, selenide, and telluride in 1940. Germanium-based bipolar transistor discovery in 1949 and the development of the silicon-assisted field-effect transistor in the late 1950s were pivotal events in

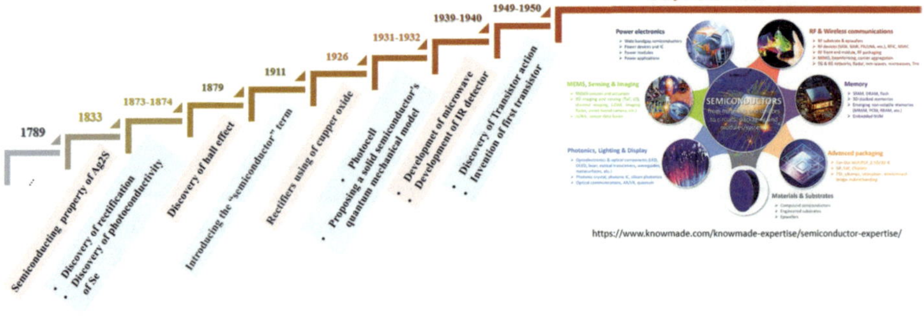

Figure 1.
A schematic brief of the history of semiconductors.

semiconductor physics. The introduction of silicon marked the beginning of semiconductor microelectronics' progress. Later, compounds including elements from the periodic table's III–V groups, such as GaAs, contributed similarly to the growth of semiconductor optoelectronics [1, 3].

Figure 1 depicts the schematic history of semiconductors from inception to the present day. Today, semiconductor-based integrated circuits, also known as microelectronic chips or computer chips, consist of billions of components capable of processing, storing, sensing, and moving data or signals. They play an important role in all industries and human existence. In this regard, considerable research and development into new semiconductor materials with improved characteristics for various applications is ongoing.

2. Semiconductor's categories

2.1 Chemical composition classification

So far, semiconductors have been categorized from different points of view. Generally, they can be either elements, compounds, or alloys, **Figure 2**.

i. Si and Ge are the most prevalent elemental semiconductors.

ii. Compound semiconductors can be produced from a variety of compositions, including II-VI, III-V, III-VI, IV-IV, and IV-VI. GaP, SiC, GaAs, GaN, and SiC are examples of compound semiconductors.

iii. Alloys comprise (a) Binary (like Sit-xGe), (b) Ternary (like AlxGat-xAs AlxGat-xN), and (c) Quaternary components (like AlxGal_, AsySbl_y GaxInt_xAs,_yP).

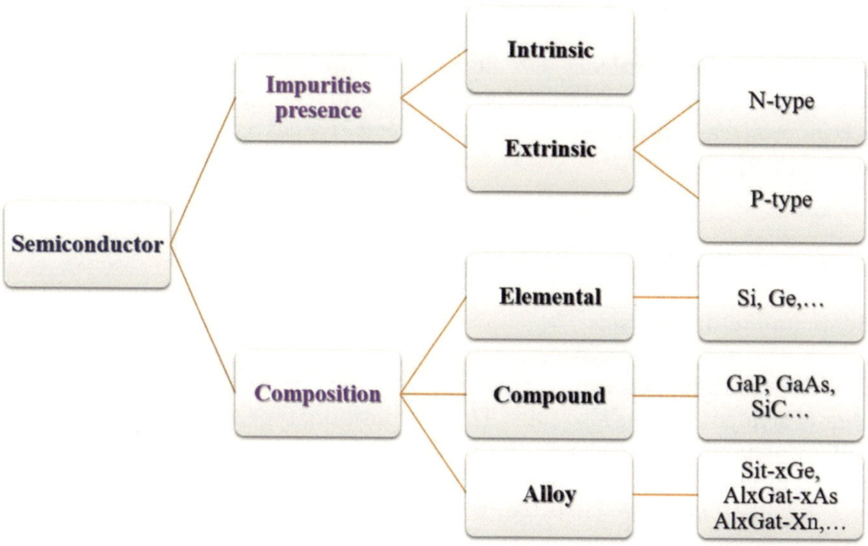

Figure 2.
Semiconductor's categories.

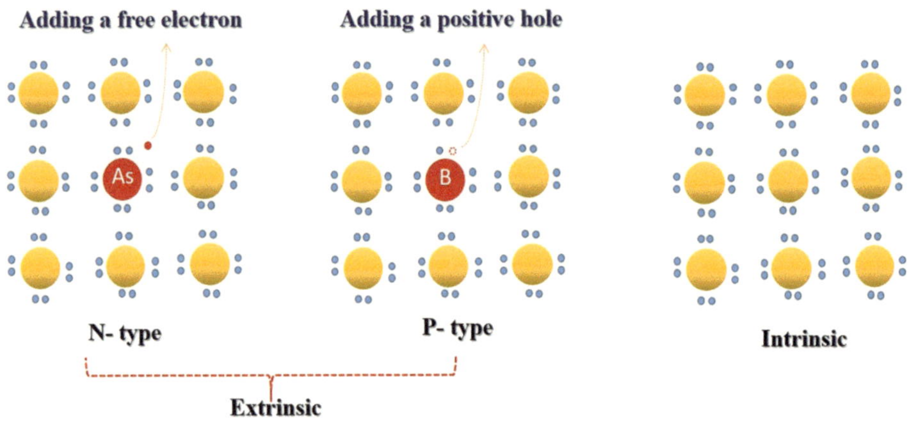

Figure 3.
Intrinsic and extrinsic semiconductors.

2.2 Intrinsic and extrinsic

Semiconductors are also broadly classified into two classes: intrinsic and extrinsic, **Figure 2**. Intrinsic semiconductors are chemically very pure and undoped. Thermal excitation or crystal defects are the reasons for poor conductivity in intrinsic semiconductors. An intrinsic semiconductor carries an equal number of negative and positive charge carriers. The quantity of charge carriers is dictated by the materials' properties, not by the amount of impurities present. Extrinsic semiconductors are made by introducing a small quantity of impurities to intrinsic semiconductors through the doping procedure, **Figure 3**. This change can alter the electrical characteristics of the semiconductor and improve conductivity. Two groups of extrinsic semiconductors, n-type and p-type, can be produced depending on the doping process. By adding valence-five elements impurities, group V of the periodic table (P, As, Sb, etc....), to valence-four semiconductors, the number of free negative charge carriers increases and n-type semiconductors (negative charge semiconductor) can be produced. Also, by adding impurities of valence-five elements, group V of the periodic table (P, As, Sb, etc....), to valence-four semiconductors, the number of free negative charge carriers is increased and n-type semiconductors (negative charge semiconductor) can be produced [6–9].

3. Semiconductor's key characteristics

3.1 Semiconductor's physics

Crystalline solids are the most typical semiconducting materials, but amorphous and liquid semiconductors also exist. The composition, type, and size of semiconductors all influence their properties. As already mentioned, in general, the physical properties of semiconductors are between conductors and insulators. For example, metallic conductivity ranges between 10^6 and 10^4 (ohm-cm)$^{-1}$. However, common insulators exhibit conductivities of less than 10^{-10} (ohm-cm)$^{-1}$. The conductivity of semiconductors ranges from 10^6 to 10^{-10}(ohm-cm)$^{-1}$. **Table 1** summarizes some characteristics of semiconductors [10–12].

Characteristics	Semiconductor
Conductivity	Moderate
Resistivity	Moderate
Forbidden gap	Small forbidden gap $\Delta E \leq 3.2$ eV (at T = 0 K)
Temperature Coefficient	Negative
Conduction	Very small number of electrons for conduction
Conductivity value	10^{-7} mho/m to 10^{-13} mho/m, distinguishing conductors from insulators
Resistivity value	10^{-5} Ω-m to 10^{5}Ω-m, classifying them as conductors or insulators
Current flow	Due to holes and unbound electrons
Number of carriers at normal temperature	Low
Zero Kelvin behavior	Acts like an insulator
Formation	Formed by covalent bonding
Valence electrons	The outermost shell contains a total of four valence electrons

Source: Reprinted from Elsevier 2023.

Table 1.
Semiconductors properties [10].

3.2 Band structure

By developing Bloch's theory and concept of electrons in crystal lattices, Brillouin offered the idea of forbidden gaps and adapted it to actual solids. After developing the periodic potential model, the solid band theory was established based on the concept of empty and full energy bands. In this regard, when atoms come into close contact to create a crystal, they interact and form energy bands. These energy bands are permissible energy states of an electron moving in the periodic potential of crystalline lattices. Among them, the upmost energy band containing electrons at temperatures higher than absolute zero is the conduction band (CB). The next lower band, separated from the conduction band by an energy gap of the order of an eV in semiconductors, is called the valence band (VB), which is mainly occupied by electrons at the desired temperatures. The energy levels in the conduction band and the unoccupied states in the valence band are occupied by electrons and holes, respectively [2, 3, 13]. But how are valence and conduction bands formed?

Band theory of solids is a theoretical model that describes the quantum state of electrons in solid materials. Following Pauli's exclusion principle, electrons in atoms are placed in their respective energy orbits. When two atomic orbitals join, they create a molecular orbit with two unique energy levels. A band is formed in solids by 10^{23} stacked lines restricted to a narrow region. Band theory aids in visualizing the distinctions between a conductor, semiconductor, and insulator by illustrating the possible energies of electrons in a substance. Consider an atom with n electrons. These electrons fill the energy level using Pauli's exclusion principle. To obey Pauli's exclusion principle, when atoms are brought together to form molecules, they must have different configurations than when they were separate atoms. In practice, their various energy bands overlap to form something called a molecular orbital. Three atomic orbitals overlap once the third atom is added to the mix, generating an individual molecule orbital with three different energy levels. Each molecular orbital will have three energy levels. In general, as additional atoms are added, the molecular orbit's energy levels increase. Eventually, we will have a whole solid made up of something like 10^{23} atoms stacked together. Now each molecular orbital will have 10^{23} distinct energy states. As a result, the gaps between them will be so narrow that individual energy levels will no longer be noticeable. Therefore, it is more convenient to imagine it as a continuous energy. Now it can be referred to as an energy band rather than a molecular orbit. In such a system with n atoms, each 1 s orbital or 2 s orbital can accommodate two electrons apiece. As a result, the total number of electrons that a 1 s and 2 s energy band can accommodate is 2n. An individual 2p state can fit six electrons, hence a 2p energy band can hold 6n electrons, and so on. **Figure 4** depicts how the energy bands of a solid made up of n atoms are formed by the change and evolution of the energy states of a single atom.

Valence electrons are those found in the outermost shell. These electrons contain a range of energy levels and form the valence band, which has the greatest occupied energy. Due to the valence electrons' weak bond with the nucleus, some of them can escape from the outermost orbit and form free electrons even at ambient temperature. These electrons, known as conduction electrons, are responsible for conducting current in conductors. The conduction band includes conduction electrons and has the fewest number of filled energy states. The band gap, or the forbidden gap, is the region between the conduction and valence bands. As the name implies, the forbidden gap has no energy, and no electrons remain in it. In this regard, if the energy gap $\Delta E \leq 3.2$ eV (at T = 0 K), the material is a semiconductor, and for $\Delta E > 3.2$ eV, the material is an insulator. This 3.2 eV is the energy available from sunlight.

It is worth noting that materials with a small overlap in the valence and conduction bands are called semimetals. Semimetals, unlike regular metals, have both kinds of charge carriers, but in smaller quantities than real metals. The electrical characteristics of semimetals lie between those of metals and semiconductors. The band structure of semiconductors, metals, insulators, and semimetals is depicted in **Figure 5**.

3.2.1 Effect of doping on the band structure of semiconductors

As mentioned, an extrinsic semiconductor (n or p semiconductors) is created via doping the crystal lattice with a small fraction of another type of atom. As a result, the

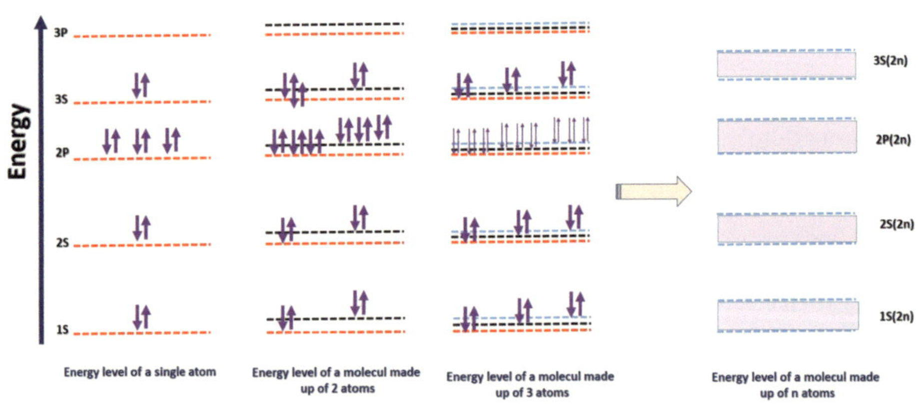

Figure 4.
Energy band development in a material containing n atoms.

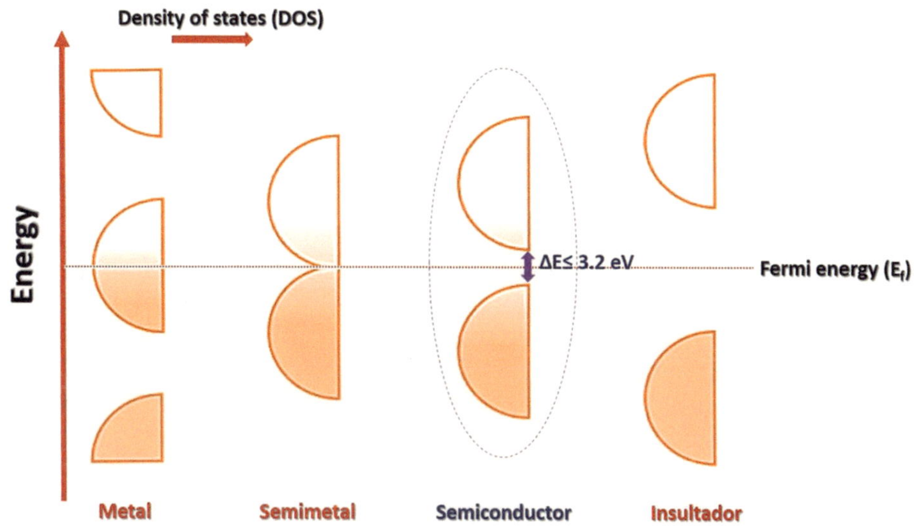

Figure 5.
The band structure of semiconductors, metals, insulators, and semimetals.

extra carriers are introduced to the structure. However, these additional carriers have a different energy structure than the lattice they are incorporated into. In n-type semiconductors, the extra carriers are electrons. These electrons have substantially higher energy than the valence electrons of the host lattice, resulting in a donor level in the band gap. The donor levels are frequently extremely near to the conduction band; therefore, thermal energy is adequate to excite these extra carriers into the conduction band. In p-type semiconductors, the dopant atoms create extra holes in the conduction band. These holes create an *acceptor level* near the valence band. The holes are readily stimulated to the valence band and conduct. **Figure 6** depicts the changes to the energy structure and the energy gap by the introduction of extra carriers [14–17].

Band engineering is a key strategy to control and alter physical, electrical, and optical properties and subsequent applications of semiconductors. In this regard, in recent years, numerous efforts have been conducted to investigate the effect of altering the band structure on the characteristics of semiconductor materials.

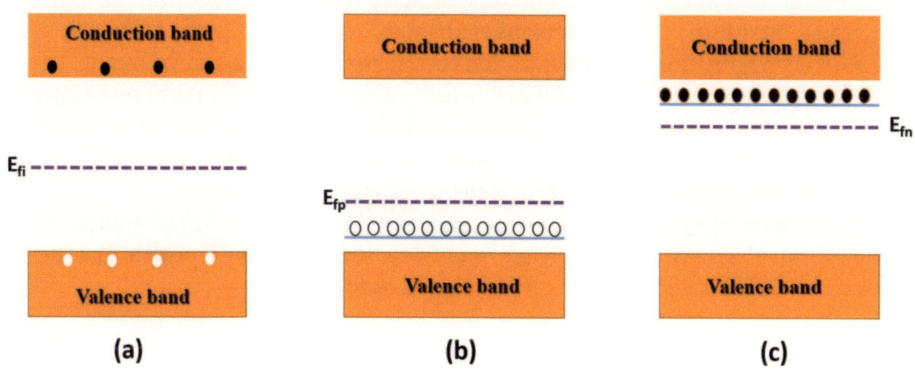

Figure 6.
Energy bands of (a) intrinsic semiconductor, (b) extrinsic p-type semiconductor, and (c) extrinsic n-type semiconductor.

For example, doping of semiconductor phosphors with rare earth (RE) ions provides unique routes to induce new emission centers, enable tunable luminescence, and improve sensing efficiency [18–21]. We recently have shown that the incorporation of RE ions into the structure through the adjustment of the lattice symmetry results in enhanced scintillation and photoluminescence characteristics in lead tungstate nanoparticles [22]. In another study, high light yield, persistent, and thermo-stimulated luminescence is obtained by defect engineering in co-doped $LiLuF_4$:Tb, Cu microcrystalline scintillation material [23]. In photocatalyst materials, doping is considered a practical approach for energy gap engineering, regulating the optoelectronic properties, and extending the performance under visible light irradiation [24–27]. In this regard, Kanoun et al. [28] have demonstrated that the photocatalytic performance of visible light in Au and Au/N doped anatase TiO_2 is enhanced when dopant ions are substituted in the TiO_2 lattice by decreasing the band gap. In another study, highly efficient visible light photocatalytic performance was achieved by the deposition of Ag nanoparticles on $AgIng_5S_8$ [29]. In addition, conductivity and magnetization are significantly altered as a result of the doping process. For instance, incorporating Fe ions into the $BaZrO_3$ or CeO_2 structures leads to an extraordinary p-type conductivity or Co-doped $KNbO_3$ perovskite exhibits ferromagnetic properties at room temperature [30–32].

3.3 Direct and indirect band gap semiconductors

In direct gap intrinsic semiconductors like Gallium arsenide, the maximum energy of the valence band and the minimum energy of the conduction band coincide at the same "K" point. In contrast, in indirect gap semiconductors like Si or Ge, the minimum energy of the conduction band and the maximum energy of the valence band occur at distinct "K" points.

3.4 Electrical conductivity

In 1914, Johann Koenigsberger categorized solid-state materials based on conductivity into three groups: insulators, metals, and variable conductors which today are categorized into the four major groups of semiconductors, metals, insulators, and semimetals [2]. The physical properties of each of these categories can be explained using energy band structures [33].

The Fermi level is a significant outcome of band theory, representing the highest occupied quantum state of electrons at absolute zero temperature, **Figure 6**. The Fermi level's position in relation to the conduction band is a crucial factor that influences the electrical characteristic of semiconductors. The electrical resistivity of a semiconductor is situated between that of a conductor and an insulator, specifically ranging from 10^3 to 10^{-8} Siemens/cm. Applying an external electric field can alter the resistance of the semiconductors. While electrons are responsible for carrying electric current in conductors, electrons or holes carry electric current in semiconductors.

The overall electrical conductivity, which is caused by the superposition of the conductivity of holes and electrons, is expressed as follows:

$$\sigma = n_e q_e \mu_e + n_h q_h \mu_h \tag{1}$$

where the electron and hole's concentration, charge, and mobility are presented by n_e, q_e, μ_e, n_h, q_h, and μ_h, respectively. Mobility is a parameter that connects the applied electric field E across the substance to the drift velocity of electrons. In the intrinsic semiconductors, n_e is equal to n_h. According to Eq. (1),

$$\sigma = n_e q_e (\mu_e + \mu_h) \tag{2}$$

$$n_e = N_c \exp(-(E_C - E_f) / k_B T, \tag{3}$$

$$n_h = N_v \exp(-(E_f - E_v) / k_B T, \tag{4}$$

where,

$$N_c = 2(m_e^* k_B T / 2\pi\hbar)^{3/2}, \tag{5}$$

$$N_v = 2(m_h^* k_B T / 2\pi\hbar^2)^{3/2} \tag{6}$$

where N_c and N_v represent the effective density of states of the valence band and conduction band edges, respectively.

In extrinsic semiconductors, the dopant concentrations, n_e and n_h, at room temperatures are higher than the concentrations of intrinsic carriers, n_i, which are generated thermally. The carrier concentrations and mobility are what determine conductivity. Therefore, while calculating the temperature-dependent conductivity, the effects of temperature on carrier concentration and mobility should be taken into account separately. The electrical conductivity of semiconductors, which can also be altered by doping them with impurities, enables the development of sophisticated electronic components that drive today's digital world [34]. In this regard, numerous investigations into enhancing the n- and p-type conductivities of semiconductors have been carried out [35–38]. For instance, Baturay et al. [39] have shown that metal doping can enhance the n-type conductivity of CuO thin films. In another study, the p-type doping impact via Nb doping is highlighted by Li et al. [40], which demonstrated that Nb doping may reduce n-tye conductivity in MoS_2 monolayer and exhibit ambipolar transport behavior under the sulfur atmosphere.

4. Application of semiconductors

Semiconductors are extremely essential for a variety of applications due to their easily regulated features. They have revolutionarily changed today's life by enabling more small, affordable, reliable, and fast electronic gadgets like LED TVs, DVD players, computers, cell phones, etc. In short, the application of semiconductors can be classified into energy, electronics, sensing, biomedical, antibacterial agents, anticancer behavior, quantum applications, water treatment, etc.

4.1 Application in electronics

Semiconductors play a critical role in modern electronics due to their distinctive electrical characteristics, which allow precise control over charge carriers. These materials form the foundation of key electronic components such as transistors, diodes, and integrated circuits (ICs). Transistors, particularly metal-oxide-semiconductor field-effect transistors (MOSFETs), are integral to digital logic systems and act as switches and amplifiers in computing and communication devices. Semiconductor diodes are essential for rectification, signal modulation, and power regulation. Furthermore, semiconductor-based ICs enable the miniaturization and integration of complex electronic functions in applications such as data processing, memory storage, and sensor technology, making semiconductors indispensable in advanced electronic systems [41, 42].

4.2 Application in energy

The application of semiconductors in energy can be divided into two parts: energy conversion and storage. Semiconductor materials, such as silicon, cadmium, CdTe, and GaAs, are widely used in photovoltaic cells for converting sunlight into electricity. Innovations in thin-film semiconductors and perovskite-based materials are also improving the efficiency and reducing the cost of solar energy systems, making them a key player in sustainable energy generation.

Semiconductors also play a crucial role in energy storage technologies, particularly in batteries. Lithium-ion batteries, used in electric vehicles and portable electronics, utilize semiconductor materials in their electrodes, improving charge capacity and cycle life. Research into semiconductor-based materials, like silicon anodes, is helping to develop next-generation batteries with higher energy density and longer lifespans [12, 41–43].

4.3 Semiconductor application in biomedical

Owing to their distinct thermal, optical, and electrical characteristics, semiconductor materials are increasingly utilized in biomedical applications specifically in developing antibacterial agents, anticancer applications, and medical imaging. Semiconductor nanomaterials like ZnO and TiO_2, have shown strong antibacterial properties. In biosensing, semiconductor-based field-effect transistors are key for detecting biomolecules due to their sensitivity to changes in surface charge, enabling real-time and highly sensitive detection in clinical settings. Additionally, semiconductors like quantum dots are used in bioimaging due to their tunable fluorescence, allowing for enhanced visualization in imaging techniques. Furthermore, their integration into drug delivery systems, such as targeted cancer therapies, is also prominent, where semiconductor nanoparticles can act as carriers or triggers for drug release through external stimuli like light or heat. Their surface can also be functionalized by targeting molecules to deliver chemotherapy drugs specifically to cancer cells, minimizing damage to healthy tissues [22, 41, 42, 44, 45].

4.4 Environmental application

Semiconductors are pivotal in environmental applications, particularly in energy conversion, pollution mitigation, and environmental monitoring. In photovoltaics, semiconductor materials like silicon and cadmium telluride are employed to convert solar energy into electrical energy through the photovoltaic effect, contributing to renewable energy technologies. Photocatalytic semiconductors, such as TiO_2, are used in advanced oxidation processes to degrade organic pollutants in air and water, leveraging electron-hole pair generation under UV or visible light for effective pollutant decomposition. Additionally, semiconductor-based gas sensors, often utilizing metal oxides like ZnO or SnO_2, are widely used for detecting hazardous gases and monitoring environmental pollutants with high sensitivity. These technologies underscore the critical role of semiconductors in advancing sustainable environmental solutions.

Among all the mentioned applications, the use of photocatalyst semiconductors for water purification and energy production has received considerable attention due to energy crises and the shortage of freshwater resources. Perhaps the first explanation of the photocatalytic capabilities of several metal oxides including ZnO, Sb_2O_3, and TiO_2 was presented around 1955 by Markham, who dealt with a variety of photochemical alterations that oxides may experience, such as accelerated oxidation of organic substances exposed to UV light. In 1972, Fujishima and Honda achieved a revolutionary finding by revealing the ability of a TiO_2 electrode to split water and produce hydrogen [46]. Following that, there was an increasing drive to research, develop, and introduce new photocatalyst materials and improve their performance. Since then, various types of semiconductors including metal oxides, sulfides, nitrides, and organic-inorganic hybrids were introduced for photocatalysis applications and photocatalysis has gained increasing interest because of its environmentally friendly procedure and straightforward conversion of solar energy into conveniently stored hydrogen. To improve the efficacy of the photocatalytic process, several techniques have been developed during the last decades to move the commencement of absorption to longer wavelengths. In this regard, doping the semiconductor photocatalyst with anions, cations, and metal ions was the first approach. Another possibility could be to use heterojunctions made up of two low band gap semiconductors with the adequate positioning of corresponding CB and VB bands via the quantum size effect. The first semiconductor heterostructures were successfully produced in the 1960s, and they quickly found uses in optoelectronic devices and electronic systems [47]. In 1984, Serpone et al. [48] demonstrated that combining CdS and TiO_2 resulted in considerable increases in the photocatalytic process. Since then, various types of heterostructured semiconductors have been developed, which can be classified into three main groups according to the following categories [49–54].

4.4.1 Different types of heterostructures

4.4.1.1 Type-II heterostructures

In type-II heterostructure, after irradiation and producing electron and hole pairs in photocatalysts I and II, photogenerated electrons were moved from photocatalysts I to photocatalysts II, and photogenerated holes were transferred oppositely. Consequently, photocatalysts II and I will accumulate photogenerated electrons and holes for the reduction and oxidation reactions, respectively. Photogenerated electrons and holes separate as a result of this process. However, the claimed charge-transfer mode in type-II heterojunction is questionable because of its inability in photogenerated electrons and holes separation. In addition, the total redox ability of the heterojunction is decreased by the transfer mode. The ongoing transfer of electrons from photocatalyst I will be prevented by the repulsion of the preexisting electrons in photocatalysts II. Moreover, the achievement of spatial charge separation is inhibited by the electric repulsion of existing holes in photocatalysis I and transferred holes from photocatalysis II. At the same time, electrostatic attraction between electrons and holes in photocatalysts I prevents electrons from moving from photocatalysts I to photocatalysts II. This makes it challenging to accomplish the intended charge-transfer process in type-II heterostructures. Furthermore, a greater charge separation rate comes at the expense of lower redox ability, which is undesirable for photocatalytic reactions.

4.4.1.2 Z-scheme heterostructures

Z-scheme photocatalysts could maintain strong redox characteristics while increasing the efficiency of charge separation. Electron donors (D) and photogenerated holes in the VB of photocatalysts I react to create the corresponding electronic acceptors (A). While D is produced as a result of the reaction of photogenerated electrons of CB photocatalyst II with A. Subsequently, the reserved holes in photocatalyst II's VB and the photogenerated electrons in photocatalyst I's CB participate in the oxidation and reduction reactions, respectively. The system may benefit from both spatially separated redox reaction sites and strong redox capabilities thanks to this charge-transfer mode. Nevertheless, there are a few shortcomings with this structure: (1) it is limited to the solution phase, (2) side reactions, (3) light shielding, and (4) pH sensitivity [55].

4.4.1.3 S-scheme heterostructures

In an S-scheme heterostructure, the recombination of ineffective charge carriers leads to a high redox potential, while strong photogenerated electrons and holes accumulate in the CB of reduction photocatalysts (RP) and the VB of oxidation photocatalysts (OP), respectively. **Figure 7** depicts how charges are transferred in an S-scheme heterostructure, which can be interpreted as follows [54]:

By contacting these two semiconductors, electrons from the RP migrate to the OP, forming two electron depletion and accumulation layers near the interface in the RP and OP, respectively. As a result, an internal electric field will form that directs electrons from the RP to the OP. This electric field boosts photogenerated electrons' transition from OP to RP instead of in the opposite direction.

When OP and RP connect, their Fermi levels shift upward and downward, respectively. Band bending led to the recombining of photogenerated electrons in OP's CB with holes in RP's VB at the interface area.

Because of Coulombic attraction, photogenerated electrons of OP and holes of RP tend to recombine at the interface.

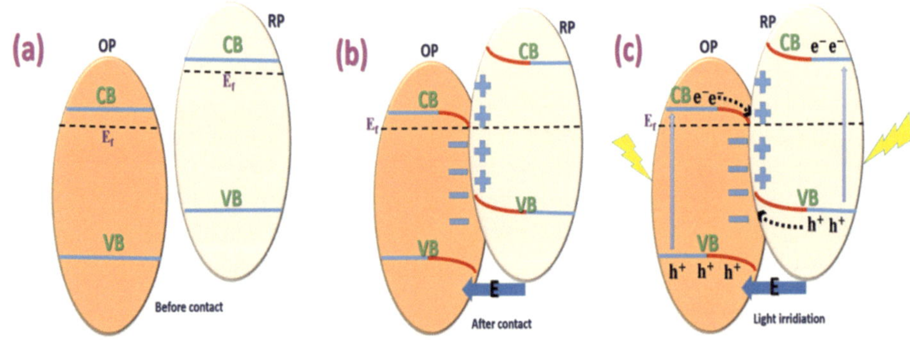

Figure 7.
Charge-transfer mechanism in an S-Scheme heterostructure (a, b) before and after contact and (c) after light irradiation [54].

4.5 Other Applications

Sensing, quantum applications, telecommunications, wireless communications, photovoltaic cells, optoelectronics, etc. are among the other applications of semiconductors. These and numerous more domains of modern and future life require the discovery of new generations of semiconductor materials that can operate faster, with more accuracy, high sensitivity, and resolution.

5. Manufacturing process

The process of transforming raw materials into finished items is known as manufacturing. Semiconductor manufacturing is a complex procedure involving many laborious steps, including photolithography, etching, doping, and testing. This process is a resource-intensive, high-stakes industry with minimal margin for error. Metals, semiconductor materials, dopants, and insulators are examples of input materials used in this process. The resulting products are printed circuit boards, integrated circuits (ICs), IC packages, and different types of commercial electronic devices and systems, such as TVs, computers, smartphones, cameras, and so on. The procedures that are used in the production of semiconductors include oxidation, diffusion, etching, planarization, ion implantation, photolithography, and crystal growth [56]. Today's expanding applications that require high speed and processing power have driven the demand for high-performance semiconductors. Thus, there is a greater need than ever for research and development as well as the application of cutting-edge technologies like artificial intelligence in manufacturing processes to optimize material usage, lower costs and defects, and boost production efficiency and quality [57–59].

6. Semiconductor devices

Today, almost every commercial product we come into contact with includes semiconductors, and semiconductor devices are everywhere. The electronic industry is built on semiconductor devices because semiconductors have easily tunable characteristics and can function as both conductors and insulators. These make them practical and beneficial in the electronics industry. Most electronic devices can be constructed from a set of building blocks as shown in **Figure 8**. P-N junctions, which are the key compound for numerous semiconductor devices, are the first building blocks, **Figure 8**(a). A P-N junction, the base of two-terminal semiconductor devices, is a junction of p-type and n-type semiconductors. The three-terminal semiconductor, the P-N-P bipolar transistor, can be produced by combining two p-n junctions. Also, combining three P-N junctions can form a switching device called a thyristor with a p-n-p-n structure. The metal-semiconductor interface, **Figure 8**(b), is the second building block. This interface can be used as a rectifying contact, i.e., the device allows current in one direction, as in ohmic contact. Metal-semiconductor field-effect transistors can be formed by using rectifying contacts as gates. The third building block is formed between two dissimilar semiconductors like GaAs and AlAs by band gap engineering, **Figure 8**(c). The metal-oxide-semiconductor structure (MOS), a combination of a metal-oxide and an oxide-semiconductor interfaced, is the fourth building block, **Figure 8**(d).

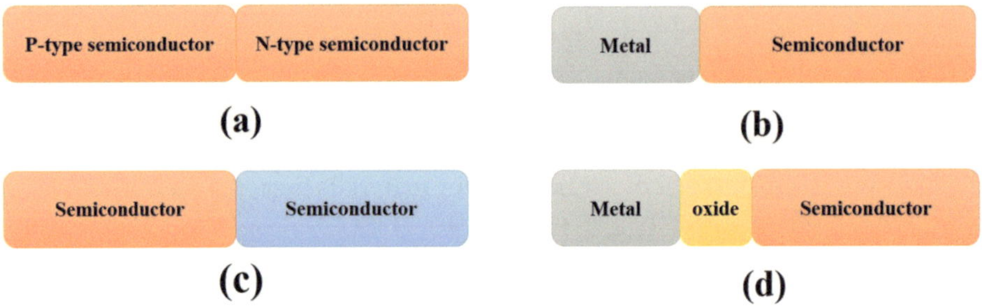

Figure 8.
Four building blocks of semiconductor devices.

This structure is used as a gate and the two semiconductor-metal oxide junctions are the source and drain. The result is the MOS field-effect transistor (MOSFET). The MOSFET is the most important component of modern integrated circuits, enabling the integration of millions of devices per chip [60].

7. Nanotechnology and the future trends of semiconductors

To stay ahead of the curve in innovation and adjust to changing industry demands, it is essential to forecast the future of semiconductor technology. One significant trend in semiconductor technology is Moore's Law, which states that transistor density will double approximately every two years. This law has been a major factor in the expansion of the semiconductor industry for many years. However, it gets harder to keep up this rate of miniaturization as transistor sizes get closer to their physical limits. Therefore, today's research efforts are focused on novel materials, device configurations, and manufacturing processes that offer better performance and scalability to overcome this challenge. Nanomaterials, with their numerous benefits, are crucial for enhancing semiconductor characteristics and addressing technological demands. Novel nanomaterials such as graphene, carbon nanotubes, and two-dimensional materials have the potential to enhance semiconductor technology beyond silicon and allow for further transistor scaling. Neuromorphic computing, quantum computing, quantum communication, Internet of Things (IoT) devices, and many other cutting-edge technologies require affordable and efficient semiconductor components that can process and transfer data more efficiently.

These demands can be met by nanoscale semiconductors because of their benefits, which include stability, conductivity, and a high surface-to-volume ratio. The intricate connection between a particle's physical properties and geometry makes it more important to accurately regulate the formation of nanostructures with particular sizes, shapes, and densities. So, one of the primary advantages of nanomaterials is the capacity to enhance chemical reactions or physical interactions at the surface or contact, which significantly impacts their performance. Surface chemistry, energy, and specific surface area are all critical elements that significantly affect energy-related activities. These variables influence the kinetics and thermodynamics of reactions occurring at the interface, plus the onset and progression of phase changes. Furthermore, nanomaterials' reduced size enables more efficient charge or heat transport, as well as dimensional changes caused by specific chemical processes and phase transitions. The advancement of semiconductor technology from the sub-micron scale to the nanoscale has led to a transformative shift in product design, establishing a substantial market for semiconductors at the beginning of the twenty-first century [10].

Today, the outstanding characteristics of semiconductor nanomaterials have sparked significant anticipation and excitement in a broad variety of sophisticated technologies, including nanoelectronics, imaging devices, sensors, nonlinear and nano optics, detectors, energy conversion, solar cells, catalysis, water treatment, and so on [43, 45, 61–65].

Applying nanoscale principles has the potential for future study, advances, and widespread use of semiconductor materials. These technical innovations will have a considerable impact on semiconductor industries' performance.

8. Conclusion and future perspectives

Despite its lengthy and complex history, semiconductors have revolutionized many fields, including electronics, energy, and environmental management, beyond all expectations. Almost all of the daily appliances we use contain semiconductors. The demand for smaller, faster, and more efficient devices has always been the driving force behind the discovery of next-generation semiconductors. Today, fields such as telecommunications require higher data speeds and network capacity. Medical imaging, diagnostics, and treatment require technologies that can provide high sensitivity, resolution, and accuracy. Environmental monitoring, water treatment, pollution control, and renewable energy are other fundamental areas that can create significant changes in the future of mankind by introducing new materials. As a result, the field of semiconductor production is expanding and becoming more significant. In this regard, in recent years, there has been a surge in interest in semiconductor nanomaterials due to the notable benefits these materials have over their conventional equivalents. Nanostructured materials, especially semiconductor photocatalysts, have grown rapidly as a result of advances in synthesis techniques and new applications for solar energy and environmental remediation. The revolutionary development of semiconductors can be attributed to the application of nanotechnology principles to control the size, shape, and physical, chemical, and mechanical properties of semiconductor materials, as well as the ability to control their conductivity through the doping process.

In particular, heterostructure semiconductors are widely developed and used today in many areas of human civilization, such as light-emitting diodes, bipolar transistors, and low-noise high-electron-mobility transistors for high-frequency applications like satellite television, etc. Besides these astounding accomplishments, the application of these structures in the photocatalysis process and water splitting for hydrogen production is fascinating and promising. In particular, the heterostructured-based photocatalysis process has received considerable attention because of its environmentally friendly nature and direct process of converting solar energy into easily stored hydrogen. However, although significant progress has been made in the laboratory or research scales, different kinds of heterostructured photocatalysis have been developed and investigated. However, it is still in its early stages, and more global research is required for its commercialization and practical applications.

Acknowledgements

This research is supported by a postdoc grant from Semnan University (32231).

Author details

Maryam Hosseinpour[1,2]* and Hassan Abdoos[1]

1 Department of Nanotechnology, Faculty of New Sciences and Technologies, Semnan University, Semnan, Iran

2 Instituto de Ciencia de Materiales de Madrid (ICMM), Consejo Superior de Investigaciones Cientificas (CSIC), C/Sor Juana Inés de La Cruz, Madrid, Spain

*Address all correspondence to: hosseinpour.maryam@semnan.ac.ir; maryam.hosseinpour@icmm.csic.es

References

[1] Fu Y, Fu Y. Semiconductor materials. In: Physical Models of Semiconductor Quantum Devices. Dordrecht, Netherlands: Springer; 2014. pp. 1-66. DOI: 10.1007/978-94-007-7174-1

[2] Busch G. Early history of the physics and chemistry of semiconductors-from doubts to fact in a hundred years. European Journal of Physics. 1989;**10**:254

[3] Pearson GL, Brattain WH. History of semiconductor research. Proceedings of the IRE. 1955;**43**:1794-1806

[4] Orton JW. Semiconductors and the Information Revolution: Magic Crystals that Made IT Happen. Netherlands: Academic Press; 2009

[5] Wilson AH. The theory of electronic semi-conductors. Proceedings of the Royal Society of London. Series A, Containing Papers of a Mathematical and Physical Character. 1931;**133**:458-491

[6] Tewksbury SK. Semiconductor materials. In: Microelectronics. England: CRC Press; 2018. p. 1

[7] Shockley W. The theory of p-n junctions in semiconductors and p-n junction transistors. Bell System Technical Journal. 1949;**28**:435-489

[8] Morin FJ. Semiconductors. Digest of Literature on Dielectrics. Washington, DC, USA: IEEE; 1949. pp. 58-71. [14 April 2016]

[9] Yu P. Fundamentals of Semiconductors. Berlin: Springer; 2005

[10] Hossain N, Mobarak MH, Mimona MA, Islam MA, Hossain A, Zohura FT, et al. Advances and significances of nanoparticles in semiconductor applications–A review. Results in Engineering. 2023;**19**:101347

[11] Majumder MK, Kumbhare VR, Japa A, Kaushik BK. Introduction to Microelectronics to Nanoelectronics: Design and Technology. England: CRC Press; 2020

[12] Rahman MA. A review on semiconductors including applications and temperature effects in semiconductors. American Scientific Research Journal for Engineering, Technology, and Sciences. 2014;7:50-70

[13] Łukasiak L, Jakubowski A. History of semiconductors. Journal of Telecommunications and Information Technology. 2010;**1**:3-9

[14] Jain VK. Semiconductors. In: Solid State Phys. Cham, Switzerland: Springer; 2022. pp. 295-330

[15] Balkan N, Erol A, Balkan N, Erol A. Intrinsic and extrinsic semiconductors. In: Semiconductors for Optoelectronics: Basics and Applications. Cham, Switzerland: Springer; 2021. pp. 37-78. DOI: 10.1007/978-3-319-44936-4_2

[16] Gurylev V, Gurylev V. Extrinsic defects in nanostructured semiconductors. In: Nanostructured Photocatalyst via Defect Engineering: Basic Knowledge and Recent Advances. Cham, Switzerland; 2021. pp. 319-48. DOI: 10.1007/978-3-030-81911-8_10

[17] Bott AW. Electrochemistry of semiconductors. Current Separations. 1998;**17**:87-92

[18] Naresh V, Cha P-R, Lee N. Cs2NaGdCl6: Tb3+– A highly luminescent rare-earth double Perovskite

scintillator for low-dose X-ray detection and imaging. ACS Applied Materials & Interfaces. 2024;**16**:19068-19080

[19] Ma Y, Sun P, Guo C, Wu J, Chen F, Qin K, et al. Improvement of X-ray excited luminescence properties of Mg4Ta2O9 scintillators by rare-earth doping. Journal of Luminescence. 2024;**265**:120231

[20] Song Z, Wang J, Tong Y, Zhang Y, Wang Q, Wang C, et al. Enhancement of alpha/beta ratio in NaI: Tl, 6Li neutron-gamma scintillators by rare earth co-doping. Nuclear Instruments and Methods in Physics Research Section A: Accelerators, Spectrometers, Detectors and Associated Equipment. 2024;**1064**:169451

[21] Liu Y, Han K, Zhao S, Zhou X, Gao S, Wang Y, et al. Ce 3+−doped alkaline-earth sulfide nanocrystals for X-ray scintillation imaging screens. Journal of Materials Chemistry C. 2024;**12**:3221-3227

[22] Hosseinpour M, Abdoos H, Alamdari S, Pecharromán C, Borrell A, Menéndez JL. Alpha sensing, NIR to green light emission in Er3+ doped PbWO4 nanoparticles with modification of calcination atmosphere. Journal of Alloys and Compounds. 2024;**1010**:177189

[23] Lu H, Xu X, Li Y, Feng G, Yang J, Huang X, et al. Heterovalent Co-doped LiLuF4 composite flexible films as persistent luminescence scintillator for X-ray high-resolution extension imaging. Advanced Optical Materials. 2024;**12**:2303134

[24] Gharagozlou M, Fard NE, Ghahari M, Yaraki MT. Bimetal Cu/Ni-BTC@ SiO2 metal-organic framework as high performance photocatalyst for degradation of azo dyes under visible light irradiation. Environmental Research. 2024;**256**:119229

[25] Du T, Hu C, Li Y, Pang Y, Li M, Liu B, et al. High-efficiency photocatalytic degradation of Rhodamine B and tetracycline by Z-scheme nanostructured Ag/Ag3PO4/Ti3+-TiO2 mesocrystals heterojunctions under visible light. Materials Science in Semiconductor Processing. 2024;**172**:108081

[26] Al-Zaqri N. High visible light photocatalytic performance of V and Mo doped CuO/ZnO composites: Integration between doping and pn heterojunction. Journal of Sol-Gel Science and Technology. 2024:1-13. DOI: 10.1007/s10971-024-06339-3

[27] Yousefi S, Sadeghzadeh-Attar A. Coupling effect of Fe-doped Co3O4 nanoparticles with SrTiO3 nanotubes on the high-efficiency photocatalytic activities of basic violet 16 dye degradation and H2 evolution. Inorganic Chemistry Communications. 2024;**162**:112273

[28] Kanoun MB, Ahmed F, Awada C, Jonin C, Brevet P-F. Band gap engineering of Au doping and Au–N codoping into anatase TiO2 for enhancing the visible light photocatalytic performance. International Journal of Hydrogen Energy. 2024;**51**:907-913

[29] Deng F, Zhao L, Luo X, Luo S, Dionysiou DD. Highly efficient visible-light photocatalytic performance of Ag/AgIn5S8 for degradation of tetracycline hydrochloride and treatment of real pharmaceutical industry wastewater. Chemical Engineering Journal. 2018;**333**:423-433

[30] Astudillo A, Izquierdo JL, Gómez A, Bolaños G, Morán O. Ferromagnetism at room temperature in Co-doped KNbO3

bulk samples. Journal of Magnetism and Magnetic Materials. 2015;**373**:86-89

[31] Wu Y, Li K, Yang Y, Song W, Ma Z, Chen H, et al. Investigation of Fe-substituted in BaZr0. 8Y0. 2O3-δ proton conducting oxides as cathode materials for protonic ceramics fuel cells. Journal of Alloys and Compounds. 2020;**814**:152220

[32] Soni S, Vats VS, Kumar S, Dalela B, Mishra M, Meena RS, et al. Structural, optical and magnetic properties of Fe-doped CeO 2 samples probed using X-ray photoelectron spectroscopy. Journal of Materials Science: Materials in Electronics. 2018;**29**:10141-10153

[33] Pierret RF, Neudeck GW. Advanced semiconductor fundamentals. MA: Addison-Wesley Reading; 1987

[34] Liang Z, Zhang Y, Souri M, Luo X, Boehm AM, Li R, et al. Influence of dopant size and electron affinity on the electrical conductivity and thermoelectric properties of a series of conjugated polymers. Journal of Materials Chemistry A. 2018;**6**:16495-16505

[35] Wang M, Wang M, Lin H-H, Ballabio M, Zhong H, Bonn M, et al. High-mobility semiconducting two-dimensional conjugated covalent organic frameworks with p-type doping. Journal of the American Chemical Society. 2020;**142**:21622-21627

[36] Lu Y, Yu Z-D, Liu Y, Ding Y-F, Yang C-Y, Yao Z-F, et al. The critical role of dopant cations in electrical conductivity and thermoelectric performance of n-doped polymers. Journal of the American Chemical Society. 2020;**142**:15340-15348

[37] Liu A, Zhu H, Kim M, Kim J, Noh Y. Engineering copper iodide (CuI) for multifunctional p-type transparent semiconductors and conductors. Advancement of Science. 2021;**8**:2100546

[38] Yang C-Y, Stoeckel M-A, Ruoko T-P, Wu H-Y, Liu X, Kolhe NB, et al. A high-conductivity n-type polymeric ink for printed electronics. Nature Communications. 2021;**12**:2354

[39] Baturay Ş, Tombak A, Batibay D, Ocak YS. n-type conductivity of CuO thin films by metal doping. Applied Surface Science. 2019;**477**:91-95

[40] Li M, Yao J, Wu X, Zhang S, Xing B, Niu X, et al. P-type doping in large-area monolayer MoS2 by chemical vapor deposition. ACS Applied Materials & Interfaces. 2020;**12**:6276-6282

[41] Harish V, Tewari D, Gaur M, Yadav AB, Swaroop S, Bechelany M, et al. Review on nanoparticles and nanostructured materials: Bioimaging, biosensing, drug delivery, tissue engineering, antimicrobial, and agro-food applications. Nanomaterials. 2022;**12**:457

[42] Zhang M, Adkins M, Wang Z. Recent progress on semiconductor-interface facing clinical biosensing. Sensors. 2021;**21**:3467

[43] Zhang C, Wei Y-L, Cao P-F, Lin M-C. Energy storage system: Current studies on batteries and power condition system. Renewable and Sustainable Energy Reviews. 2018;**82**:3091-3106

[44] Lucca BG, Petroni JM. Application of II–VI semiconductor-based biosensors in Nanomedicine and bioanalysis. In: Handbook of II-VI Semiconductor-Based Sensors and Radiation Detectors. Vol. 3: Sensors, Biosensors and Radiation Detectors. Cham, Switzerland: Springer; 2023. pp. 653-672

[45] Hosseinpour M, Abdoos H, Alamdari S, Menéndez JL. Flexible Nanocomposite scintillator detectors for medical applications: A review. Sensors and Actuators A: Physical. 2024;**378**:115828

[46] Fujishima A, Honda K. Electrochemical photolysis of water at a semiconductor electrode. Nature. 1972;**238**:37-38

[47] Dubrovskii VG, Cirlin GE, Ustinov VM. Semiconductor nanowhiskers: Synthesis, properties, and applications. Semiconductors. 2009;**43**:1539-1584

[48] Serpone N, Borgarello E, Grätzel M. Visible light induced generation of hydrogen from H 2 S in mixed semiconductor dispersions; improved efficiency through inter-particle electron transfer. Journal of the Chemical Society, Chemical Communications. 1984;**6**:342-344

[49] Alsalme A, Galal AH, El-Sherbeny EF, Soltan A, Abdel-Messih MF, Ahmed MA. Fabrication of S-scheme TiO2/g-C3N4 nanocomposites for generation of hydrogen gas and removal of fluorescein dye. Diamond and Related Materials. 2022;**122**:108819

[50] Guo M, Jiang W, Ding J, Lu J. Highly active and recyclable CuO/ZnO as photocatalyst for transesterification of waste cooking oil to biodiesel and the kinetics. Fuel. 2022;**315**:123254

[51] Upadhyay GK, Rajput JK, Pathak TK, Kumar V, Purohit LP. Synthesis of ZnO: TiO2 nanocomposites for photocatalyst application in visible light. Vacuum. 2019;**160**:154-163

[52] Chaudhary K, Shaheen N, Zulfiqar S, Sarwar MI, Suleman M, Agboola PO, et al. Binary WO3-ZnO nanostructures supported rGO ternary nanocomposite for visible light driven photocatalytic degradation of methylene blue. Synthetic Metals. 2020;**269**:116526

[53] Changlin YU, Kai Y, Qing SHU, Jimmy CYU, Fangfang CAO, Xin LI. Preparation of WO3/ZnO composite photocatalyst and its photocatalytic performance. Chinese Journal of Catalysis. 2011;**32**:555-565

[54] Hosseinpour M, Mirzaee O, Alamdari S, Menéndez JL, Abdoos H. Novel PWO/ZnO heterostructured nanocomposites: Synthesis, characterization, and photocatalytic performance. Journal of Environmental Management. 2023;**345**:118586

[55] Ng B, Putri LK, Kong XY, Teh YW, Pasbakhsh P, Chai S. Z-scheme photocatalytic systems for solar water splitting. Advancement of Science. 2020;**7**:1903171

[56] May GS, Spanos CJ. Fundamentals of Semiconductor Manufacturing and Process Control. United States: John Wiley & Sons; 2006

[57] Zwerver AMJ, Krähenmann T, Watson TF, Lampert L, George HC, Pillarisetty R, et al. Qubits made by advanced semiconductor manufacturing. Nature Electronics. 2022;**5**:184-190

[58] Maitra V, Su Y, Shi J. Virtual metrology in semiconductor manufacturing: Current status and future prospects. Expert Systems with Applications. 2024;**249**:123559

[59] Raghuweanshi P. Revolutionizing semiconductor design and manufacturing with AI. Journal of Knowledge Learning and Science Technology. ISSN 2959-6386. 2024;**3**(3):272-277

[60] Shaheen A, Zia W, Anwar MS. Band Structure and Electrical Conductivity in Semiconductors. Lahore, Pakistan: LUMS School of Science and Engineering; 2011

[61] Hosseinpour M, Abdoos H. Manufacturing of Nanocomposites via powder injection Molding: Focusing on thermal management systems - a review. Journal of Manufacturing Science and Engineering. 2020;**143**:1-40. DOI: 10.1115/1.4048454

[62] Hosseinpour M, Abdoos H, Mirzaee O, Alamdari S. Fabrication and characterization of a new flexible ionizing ray sensor based on lead tungstate (PbWO4). Ceramics International. 2023;**49**:4722-4732

[63] Hosseinpour M, Mirzaee O, Alamdari S, Menéndez JL, Abdoos H. Development of a novel flexible thin PWO (Er)/ZnO (Ag) nanocomposite for ionizing radiation sensing. Journal of Alloys and Compounds. 2023;**967**:171678

[64] Hosseinpour M, Abdoos H, Alamdari S, Mirzaee O, Menéndez JL. Lead tungstate/zinc oxide Heterostructure Nanocomposites: Luminescent properties under UV/red laser excitation. Journal of Composites and Compounds. 2023;**5**:xx-xx

[65] Sekar N, Ramasamy RP. Recent advances in photosynthetic energy conversion. Journal of Photochemistry and Photobiology C Photochemistry Reviews. 2015;**22**:19-33

Chapter 2

Advances in Photovoltaic Detectors: Principles, Challenges, and the Role of Avalanche Detectors in High-Sensitivity Detection

Jiyuan Zheng, Xiaoran Wang, Beichen Liu, Rui Wu, Jizhe Zhao, Yinjie Liu, Xiayang Hua, Yubo Yang, Zhibiao Hao, Anran Guo and Lai Wang

Abstract

This chapter mainly discusses the fundamental principles of photovoltaic detection, namely, the energy conversion procedure of light into electrical signals in photodetectors (PD) and avalanche photodetectors (APD). After briefly introducing the basic principles of PIN PD and APD, the chapter mainly focuses on analyzing dark current's noise sources and origin, bridging the relationship between the device's sensitivity performance and its material and device characteristics. From these analyses, the readers could understand that the sensitivity of PIN PD, especially under high data rates, can be significantly enhanced by merely increasing the bias voltage to turn it into APD mode. A step further, this chapter provides deeper insights into modifying the APD structure from the perspective of carrier transportation and impact ionization engineering. Besides, Geiger mode APD, which is capable of sensing weak light signals at a single photon level, is also described. In the final part of the chapter, the challenges of these detectors are discussed, and many recent technical approaches, including linear high-gain APD, digital alloys APD, and SPAD with dynamic memristor quenching, have been presented to tackle these challenges. By demonstrating the basic principles of photovoltaic detectors, particularly APD, to the analysis of receiver sensitivity, this chapter could provide comprehensive and valuable insights for designing and optimizing high-sensitivity detection systems. These systems find applications in various fields, including optical interconnection, optical communication, and LiDAR.

Keywords: photodiode, dark current, bandwidth, avalanche photodiode, sensitivity, Geiger mode photodiode

1. Introduction

As a crucial element in optoelectronic technology, the photovoltaic detector is a cornerstone in the information industry. From optical communications to imaging and sensing, they are indispensable in our lives, especially in the era of AI and IoT, where urgent demands for enormous data transportation and processing are coming from every aspect of society. Thus, improving photodetectors' performance is becoming a highly anticipated research point.

In this chapter, we will have a deep insight into the history and research of photovoltaic detectors. This chapter starts with a brief introduction of the photovoltaic effect and the fundamental principle of today's photodetectors in part 2, together with the discussions about different types of detectors classified by materials, structures, and operating wavelengths. The evolution of photovoltaic detectors from the simplest PN junction photodetectors (PD) to PIN PD is also included in the content. Analysis of the structure and electrical characteristics of PIN PD is given in part 3.

As a type of detector that utilizes the avalanche effect or impact ionization effect to magnify photocurrent, APD is a promising candidate for high-sensitivity and weak light detection. The mechanism of the avalanche multiplication process is discussed in part 4, elucidating the gain theory and other critical characteristics of APDs, including gain-bandwidth production, excess noise, and receiver sensitivity. The challenges and limitations are discussed in part 5. In pursuit of higher performance, several creative frontier technologies have been introduced, such as linear high-gain APDs, digital alloy APDs, and Geiger-mode APDs, which have been summarized in part 6. Their unique capabilities in ultrahigh sensitivity detection have received wide attention. In the end, conclusions are drawn in part 7.

From the basic theories, materials, and structures to the newly developed detectors, together with the analysis of optical receiver sensitivity, this chapter could provide a unique and comprehensive understanding of photovoltaic devices. The technical issues discussed are practical in multiple fields, including optical interconnections, vehicle-mounted LiDARs, intelligent city management, space communication, and quantum information science. This chapter could benefit readers in their studies and research on designing and optimizing high-quality photodetector systems.

2. Fundamental principles of photovoltaic detection

The discovery of the photovoltaic effect can be traced back to the nineteenth century. In 1839, E. Becquerel first identified the photovoltaic effect. In 1876, William Adams and Richard Day discovered the photovoltaic effect in solid selenium [1]. However, in this period, due to the lack of efficient carrier separation and collection mechanisms, as well as the low photo response of selenium material, effective detection devices could not be fabricated.

With the emergence of PN junctions and single-crystal silicon extraction technology in the 1940s and 1950s, support was provided from both structural and material perspectives, leading to the advent of photovoltaic detectors. Below, we will explore the development of photovoltaic detectors in terms of materials, structures, and their corresponding detectable optical signal wavelength.

2.1 Materials

Conventional semiconductor materials are the most widely used in photovoltaic detectors. By utilizing semiconductors with different bandgaps, optical signals of various wavelengths can be detected.

Early photovoltaic detectors primarily operated in the infrared and visible light bands, using narrow-bandgap materials such as Si, InSb, Ge, and GaAs. Si, Ge [2], and GaAs [3] are suitable for detecting near-infrared and visible light. Si, with its high strength and good thermal conductivity, is widely used, with a detection range of 400–1100 nm [4], and can also serve as a substrate material. InSb, with its extremely low bandgap of 0.17 eV, is used to detect mid-infrared signals in the 3–5 μm range [5]. For ultraviolet (UV) signal detection, wide-bandgap semiconductors are typically used, such as AlN (bandgap 6.2 eV) and GaN (bandgap 3.4 eV). These materials respond only to UV light, operate stably at high temperatures, and possess strong radiation resistance. Compared to narrow-bandgap materials that require light filters, they reduce detection costs and complexity [6].

Mixing specific semiconductor materials in proportion can form alloys with enhanced properties. Examples include InGaAs [7], HgCdTe, and AlGaN. By adjusting the ratio of In to Ga in InGaAs, its bandgap can range from 0.36 eV (InAs) to 1.42 eV (GaAs). Notably, In0.53Ga0.47As is lattice-matched with InP substrates, offering high electron mobility and excellent photo-response performance [8]. HgTe, with its very low bandgap (~0 eV), is unsuitable as a standalone detector material, but when combined with CdTe (bandgap 1.44 eV) to form HgCdTe, it can detect a wide range of infrared signals. Similarly, AlGaN, formed by mixing AlN and GaN, can detect near-ultraviolet light (200-350 nm) and some far-ultraviolet light (<200 nm). The intuitive bandgap and the corresponding cutoff detection wavelength of conventional semiconductor materials and alloy semiconductor materials are shown in **Figure 1(a)**.

Processing materials like black silicon and surface plasmonics enhances light-capturing ability by altering semiconductor surface structures. Black silicon, created via etching, features black micro-nano spikes on its surface [9], reducing reflectance (<5%) compared to traditional silicon (40%) and altering its band structure to absorb photons with energy below the bandgap, enabling strong absorption in the silicon absorption blind spot (1100–2500 nm) [10]. Surface plasmonics introduce metal nanoparticles on semiconductor surfaces, coupling with incident light to generate localized surface plasmon resonance. For instance, adding Au nanoparticles to GaAs significantly increases the absorption coefficient, with a 35% enhancement at 800 nm [11].

2.2 Structure

2.2.1 PN junction

The earliest photovoltaic detectors used PN junction structures, shown in **Figure 1(b)**, and can be classified into homojunctions and heterojunctions based on the materials used. Homojunctions, made of the same material for the P and N regions, offer advantages such as lattice matching and no interfacial defects. Heterojunctions provide greater flexibility: narrow-bandgap materials can serve as absorption layers for efficient photon absorption, while wide-bandgap materials can act as surface layers to block thermal carriers and minimize dark current and noise.

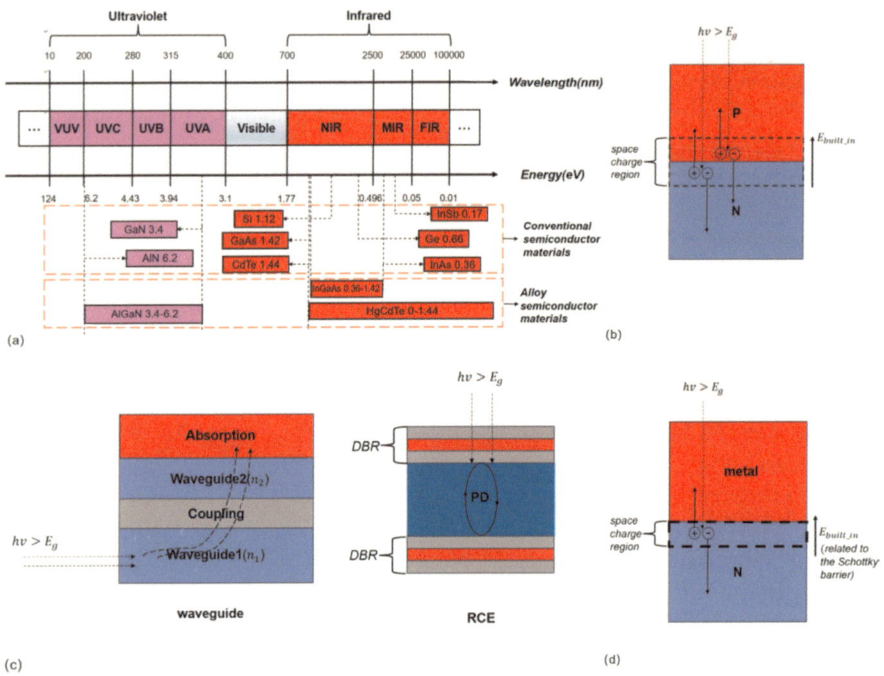

Figure 1.
(a) The bandgap and cutoff detection wavelength (maximum detection wavelength) of some conventional semiconductor materials and alloy materials. (b) Structure of the PN junction and principle of the PN junction photovoltaic detector. (c) Basis structure of waveguide and RCE photovoltaic detector. (d) Structure and basic principle of the Schottky junction photovoltaic detector.

2.2.2 PIN structure

Although PN junction detectors have a simple structure, their electric field distribution is non-uniform. When the voltage is high, the electric field at the junction can cause avalanche breakdown or tunneling, affecting the linearity and controllability of the device. Therefore, an intrinsic (i) region is inserted into the PN junction. The electric field in the i-type region changes little and can be approximated as uniform. Photons are absorbed in the i-type region, and the generated photocarriers can be transported under the uniform electric field. The linearity and controllability of the device are significantly increased.

2.2.3 Schottky structure

Schottky junction is formed by metal-semiconductor contact, commonly using Ge, Si, and graphene. The semiconductor's work function is usually smaller than the metal's. Upon contact, electrons from the semiconductor transfer to the metal until their Fermi levels align, causing an upward bending of energy levels near the semiconductor surface and forming a built-in electric field, as illustrated in **Figure 1 (d)**. Under illumination, photogenerated electron-hole pairs are separated by this field [12]. Advantages of Schottky junctions include simple structure and low dark current due to the barrier. A 2017 study on graphene/Si Schottky structures achieved a high responsivity of 0.52 A/W and a low dark current of 2.0×10^{-12} A [13].

2.2.4 Waveguide structure

To address the conflict of absorption layer thickness, optical waveguide or resonant cavity enhanced (RCE) structures can be used. As depicted in **Figure 1(c)**, the optical waveguide structure allows light to propagate through the waveguide and couple with the absorption region, improving light absorption efficiency. The RCE method involves designing distributed Bragg reflectors (DBR) at the top and bottom of the device, consisting of alternating layers of high and low refractive index materials, with each layer's thickness set to a quarter of the light's wavelength. This structure keeps the absorption layer thin while enhancing quantum efficiency through multiple reflections.

3. PIN photodetectors

3.1 Introduction of PIN photodetectors

As mentioned in 2.2.2, PIN PD is an optimized version based on PN junction detector. A PIN detector has three main functional regions: a p-type region, an i-type(undoped) region, and a n-type region. The schematic diagram of PIN PD is given in **Figure 2**. The first PIN photodiode was invented by professor Jun-Ichi Nishizawa and his research group at Tohoku University in Japan in 1952.

Because of the addition of i-type region, the electric field in the absorption region is almost uniform, greatly increasing the linearity and controllability. And, in order to prevent breakdown at the corners of the p-region caused by high electric field, a lightly doped guard ring is usually added.

The bandwidth improvement can be explained from the perspective of RC time constant. Similar to PN junctions, in PIN junctions, a positive charged layer and a negative charged layer are separated by the i-type region, which can be described as a planar capacitor. The depletion region capacitance of PIN junction can be written as [14]

$$C_{dep} = \frac{\varepsilon_s A}{W} \quad (1)$$

Compared to PN junctions, the width W of the i-type region is larger, and it does not change with applied voltage. The increased W leads to a decreased C_{dep}, resulting in a decreased RC constant and an increased bandwidth.

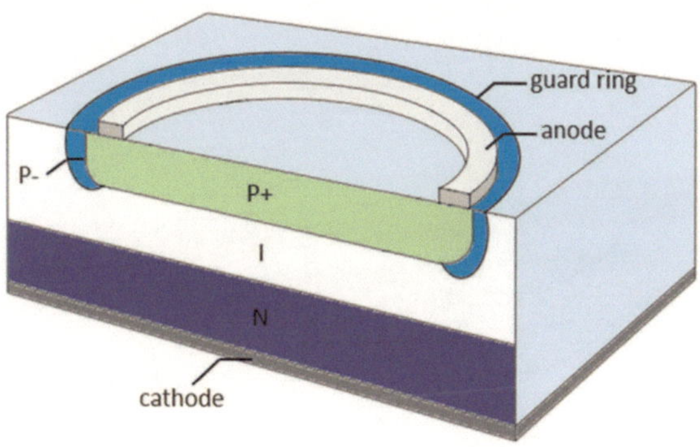

Figure 2.
Schematic diagram of a PIN detector.

3.2 Dark current and its impact on sensitivity

Dark current refers to the current flowing through a PIN detector under reverse bias without any illumination, which is one of the most critical parameters for measuring device performance.

The dark current in semiconductor PIN detectors can be divided into bulk dark current and surface dark current, which can be further divided into diffusion current, generation-recombination current, direct tunneling current, and defect-assisted tunneling current.

3.2.1 Drifting current generated by dark carriers

Drifting current refers to the current generated by the diffusion of minority carriers into the i-type region. When applying a reverse bias voltage, the electric field in the i-type region increases, enhancing the drift movement of charge carriers. Electrons near the boundary of the p-type region are driven towards the n-type region under the electric field. At the same time, holes near the boundary of the n-type region are also driven towards the p-type region, as shown in **Figure 3(a)**. When the minority carriers near these boundaries are driven away, a concentration gradient of minority carriers is formed, as shown in **Figure 3(b)**. Then, the minority carriers inside the n-type and p-type regions will diffuse towards the i-type region and then drift under the electric field in the i-type region. The sum of the two is the total reverse drifting current.

When the reverse voltage increases to a certain value, the reverse drifting current no longer increases, and the current density at this time is called the reverse saturation current density J_s. The expression for reverse current density can be written as [15]

$$J = J_s\left[\exp\left(\frac{qV}{k_0T}\right) - 1\right] = \left(\frac{qD_n n_{p0}}{L_n} + \frac{qD_p p_{n0}}{L_p}\right)\left[\exp\left(\frac{qV}{k_0T}\right) - 1\right] \tag{2}$$

Among them, V is the bias voltage, $L_p, L_n, D_p,$ and D_n are the diffusion length and diffusion coefficient of holes and electrons, respectively. p_{n0} and n_{p0} are the minority carrier concentrations in the equilibrium state of the n-type and p-type regions.

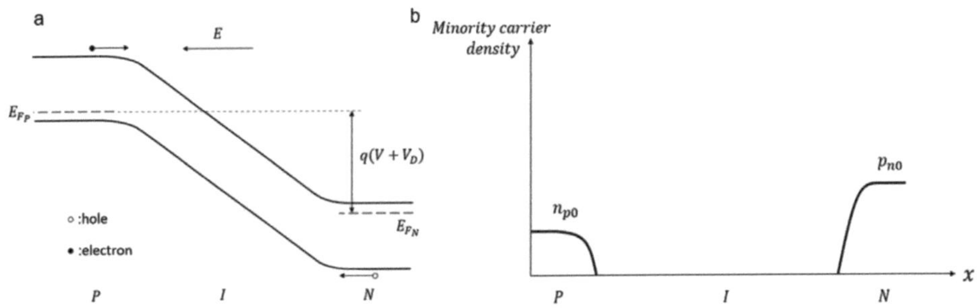

Figure 3.
Carrier transition and its dynamic distribution. (a) Electron and hole drifting in PIN junction. (b) The drifting causes the concentration gradient of minority carriers.

3.2.2 Generation-recombination current

Generation-recombination current refers to the current generated due to the excitation or recombination process of electron-hole pairs. In the thermal equilibrium state, the recombination rate is equal to the generation rate. When the PIN junction is reversely biased due to the enhanced electric field, electron-hole pairs are driven away before they can recombine, resulting in a higher generation rate and forming another part of the reverse current.

The main generation-recombination processes include Shockley Read Hall (SRH) generation-recombination, Auger generation-recombination, and direct radiation generation-recombination. The current density generated can be written as [16]

$$J_G = \frac{qn_iW}{2\tau} \tag{3}$$

And, the carrier lifetime is decided by the three processes:

$$\frac{1}{\tau} = \frac{1}{\tau_{SRH}} + \frac{1}{\tau_{Auger}} + \frac{1}{\tau_{Rad}} \tag{4}$$

SRH generation-recombination is indirect generation-recombination through a single recombination center. Let the energy level of the recombination center be E_T and the concentration be N_T. The recombination rate is [17]

$$R_{SRH} = \frac{N_T r_p r_n \left(np - n_i^2\right)}{r_n(n + n_1) + r_p(p + p_1)} \tag{5}$$

where $n_1 = N_c \exp\left(-\frac{E_c - E_T}{k_0 T}\right), p_1 = N_v \exp\left(-\frac{E_T - E_v}{k_0 T}\right)$.

Auger generation-recombination is also indirect generation-recombination. Energy generated during the recombination process excites another carrier to a higher energy level. During its transition back to a lower energy level, excess energy is released by phonon emission or exciting another electron to the conduction band.

Direct radiative recombination refers to carrier recombination caused by the direct transition of electrons accompanied by photon emission. Within a unit volume, each electron has a certain probability of recombining with a hole per unit of time. Therefore, the recombination rate is [18]

$$R = rnp \tag{6}$$

The coefficient r is only related to the velocity of charge carriers.

The reverse process of recombination is generation. In non-degenerate cases, it can be considered that the valence band is full and the conduction band is empty, so the excitation probability is independent of carrier concentration. Therefore, the generation rate is almost the same in all non-degenerate cases [18]:

$$Generation\ rate = G \tag{7}$$

At thermal equilibrium, production rate is equal to recombination rate, that is [18],

$$G = rn_0p_0 = rn_i^2 \tag{8}$$

The net direct recombination rate of carriers is [18]

$$R_{\text{Rad}} = R - G = r(np - n_i^2) \tag{9}$$

3.2.3 Tunneling current

For PIN junction, when the reverse bias voltage is high enough, the energy band is extremely bent, causing the conduction band edge of the n-type region E_{cn} to be positioned below E_{vp}, the valence band maximum of the p-type region. Electrons in the n-type region conduction band may pass through the bandgap to the p-type region valence band and vice versa.

Direct tunneling is the direct tunneling of charge carriers through the potential barrier, also known as band-to-band tunneling (BBT), as shown in process 1 in **Figure 4**. Electrons in the p-type region's valence band directly enter the n-type region's conduction band, forming a direct tunneling current.

Indirect tunneling is the process in which charge carriers tunnel through the potential barrier through a trap energy level, also known as trap-assisted tunneling (TAT), as shown in process 2 in **Figure 4**. Electrons in the valence band of the p-type region are captured by trap energy levels and then enter the conduction band of the n-type region, forming a defect-assisted tunneling current.

3.2.4 Surface dark current

Surface leakage current is caused by surface recombination, which is the recombination process through recombination centers on the surface. The number of electron-hole pairs recombined per unit surface area per unit time is called the surface

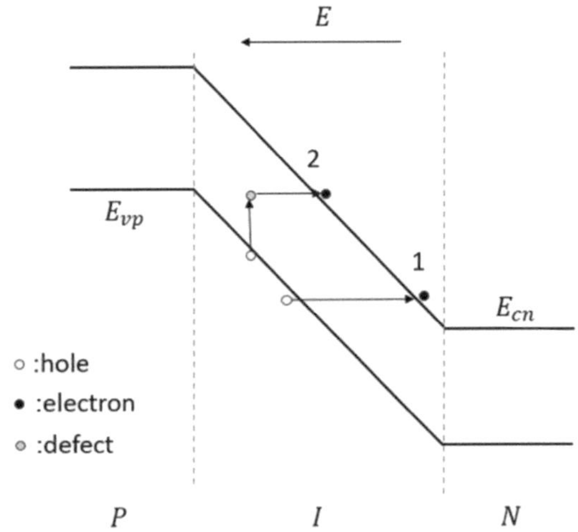

Figure 4.
Tunneling process in PIN junction.

recombination rate. The surface recombination rate U_s is proportional to the non-equilibrium carrier concentration $(\Delta p)_s$ at the surface [19]:

$$U_s = s\,(\Delta p)_s \tag{10}$$

The proportionality coefficient s has the dimension of velocity, which is called the surface composite velocity.

3.2.5 Sensitivity of PIN photodetectors

The metric usually used to quantify a photodetector's sensitivity is the noise equivalent power (NEP), which is the minimum optical power received by the detector to achieve an SNR of 1 in a 1 Hz bandwidth. Thus, the definition of NEP is [20]

$$NEP = \frac{P_{min}}{(\Delta f)^{\frac{1}{2}}} \tag{11}$$

Where P_{min} is the optical power that results in an SNR of 1 and Δf is the bandwidth of the detector. The unit of NEP is $W/\sqrt{Hz}$.

In order to show the relationship between noise and sensitivity, we can write Eq. (11) in another form [20]:

$$NEP = \frac{\left(\overline{i_n^2}\right)^{\frac{1}{2}}}{R} \tag{12}$$

Where R is responsivity of the detector and $\left(\overline{i_n^2}\right)^{\frac{1}{2}}$ is the valid value of dark noise current, measured in darkness and with $\Delta f = 1Hz$ ◦ .

The detectivity D^* is another important parameter used to describe the sensitivity of detector. It is derived from NEP [20]:

$$D^* = \frac{(A\Delta f)^{\frac{1}{2}}}{NEP} \tag{13}$$

It can also be written as [21]

$$D^* = \frac{RA^{\frac{1}{2}}}{(2qI_{dark})^{\frac{1}{2}}} \tag{14}$$

Where A is the area of the device.

3.2.6 Noise sources in PIN photodetectors

The dark noise in PIN detectors mainly includes dark current shot noise, thermal noise, and 1/f noise.

The randomness of dark current can cause fluctuations in the output signal, known as dark current shot noise. Dark current shot noise is white noise. The power of shot noise is [22]

$$\overline{i_{shot}^2} = 2qI_d\Delta f \tag{15}$$

Here, Δ_f is the working bandwidth of the detector.

1/f noise, also known as flicker noise, is a low-frequency noise whose power spectral density is inversely proportional to frequency. It is mainly caused by defects, impurities, or non-uniformity on the semiconductor surface. The expression is [23]

$$S_{1/f} = A{\cdot}f^{-\alpha} \tag{16}$$

Among them, A and α are empirical coefficients, and f is frequency.

Thermal noise is caused by the irregular thermal motion of carriers in semiconductors. It is proportional to temperature, so its impact can be reduced by cooling. Thermal noise is white noise, and its power spectral density is expressed as [22]

$$S_{thermal} = \frac{4kT}{R} \tag{17}$$

The relative contribution of these noises may vary depending on the operating conditions of the device, such as temperature and bias voltage. For example, in the low-frequency range, 1/f noise may become the main source of the noise. In the high-frequency range, thermal noise and shot noise may be more significant.

4. Avalanche photodetectors (APDs)

Traditional PIN photodetectors (PD) introduced above now have a wide range of applications; however, in the situation of distant light communication or other scenes that require the ability of weak signal detection and low energy consumption, PIN photodetectors cannot meet the sensitivity standard. Therefore, semiconductor detectors with higher sensitivity emerge as the times require, namely, the APD, which usually exhibits 5-10 dB higher sensitivity in comparison to PIN PD counterparts due to the internal gain mechanism [24], as the operating bandwidth is below 10 Gb/s. The first APD was invented by Japanese engineer Jun-Ichi Nishizawa in 1952. Typically, APDs could operate at bias voltage below or above the breakdown voltage, corresponding to linear mode and Geiger mode; APDs operating at Geiger mode are specified as single photon avalanche detectors (SPAD). As one kind of vacuum tube-based device, photomultiplier tubes (PMTs) can also amplify the weak light signal. However, PMTs are gradually being replaced since they are bulky, fragile, sensitive to the magnetic field, and need very high operating voltage.

4.1 Avalanche multiplication process and APD structure evolutions

The word Avalanche originates from French and Latin, initially referring to the spectacular phenomenon in snow mountains. An avalanche in snow mountains can be well explained by the impact-induced multiplication between falling and still snow under the gravitational field. Surprisingly, a similar phenomenon could occur in semiconductor materials as well. Just imagine the falling and still snow on hills as the free and confined carriers in semiconductors, respectively. Moreover, the dominating field is the electric field instead of the gravitational field. In detail, under a high electric field, the initial free carriers (electrons in the conduction band) will obtain enough kinetic energy after accelerations and transfer part of kinetic energy to the confined carriers (electrons in the valence band) through impact ionization as illustrated in **Figure 5**. The confined carriers then transit into free carriers, and the initial carriers remain free after the impact, and electrons and holes could participate in impact ionization. The initial and new free carriers then accelerate under the electric field and repeat the impact ionization repeatedly, which is the basic principle of avalanche multiplication, as illustrated in **Figure 6** [25].

It is noteworthy that both electrons and holes could participate in the avalanche process. The impact ionization coefficient is defined as the mean impact ionization times for a free electron or hole in unit travel distance, denoted by α and β for electrons and holes, respectively. Therefore, $\frac{1}{\alpha}$ and $\frac{1}{\beta}$ express the mean distance between two impact ionizations for free electrons and holes. Significantly, α and β are positively relevant to the electric field in the multiplication region, and the ratio between α and β is denoted by impact ionization ratio k, which is the dominating factor for APD performance of gain, bandwidth, and excess noise, k could be

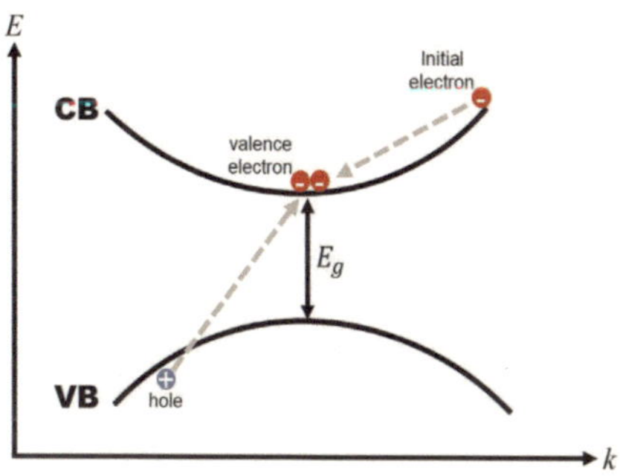

Figure 5.
The schematic diagram of the impact ionization in APD, assuming in direct bandgap material where E_g is the bandgap and only electron avalanche is presented. E is the energy axis, and k is the momentum axis; the impact ionization is denoted by gray dash arrows.

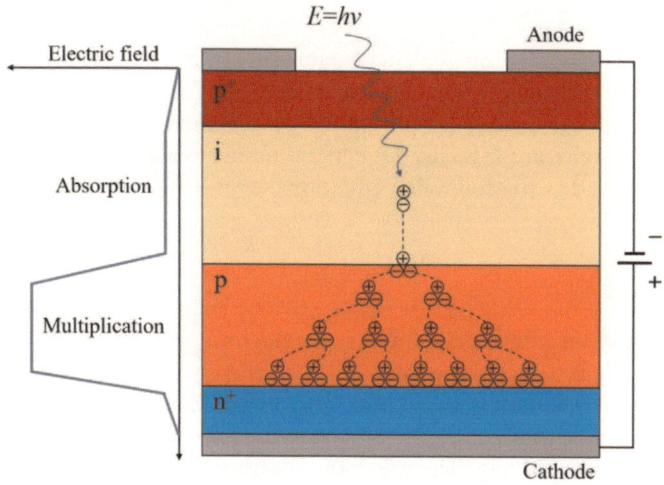

Figure 6.
Schematic diagram of avalanche multiplication of electrons. The left part denotes the electric field in the device [25]. CC BY 4.0.

expressed by either $\frac{\beta}{\alpha}$ or $\frac{\alpha}{\beta}$ as electron or holes dominates the avalanche to keep $k \in [0, 1]$ for the convenience in calculation. Typically, only one type of carrier is expected to establish an avalanche in APD to improve gain and eliminate excess noise. Hence, k is expected to be as low as possible in APD. As mentioned before, lower k also benefits avalanche gains and bandwidth, thus further contributing to better gain-bandwidth production.

For APDs, the separate absorption and multiplication (SAM) structure was proposed to eliminate the tuning dark current by adopting larger bandgap materials as multiplication regions [26]. A graded layer was introduced to avoid carrier accumulation induced by the energy stair at the interface of heterojunction, and reduce the lattice mismatch, which could form defects and thus increase dark current; in this way, separate absorption, graded, and multiplication (SAGM) APD was constructed [27].

To further improve the electric field in the multiplication region, the charge layer was designed between the absorption and multiplication layer, by tuning the thickness and the doping level of the charge layer; the electric field in multiplication region is further enhanced while the electric field in the absorption region is reduced, and an unwanted avalanche in absorption layer is suppressed while avalanche gain in multiplication region is improved [28]. Depending on whether there is a grading layer or not, APD with a charge layer design could be classified as a separate absorption, grading, charge, and multiplication (SAGCM) structure or a separate absorption, charge, and multiplication (SACM) structure. Nowadays, most III-V compound-based APDs are in SAGCM structures, while most Si or Ge on Si APDs are in SACM or SAM structures.

The discussion above could provide a good qualitative understanding of the impact ionization and the avalanche multiplication in APD, which is the origin of the internal gain and excess noise of APD. The structures of typical APD are also briefly summarized. The gain mechanism and excess noise will be discussed in detail in the next section.

4.2 Gain mechanism, gain-bandwidth production, excess noise, and receiver sensitivity

Gain in APD is specified by avalanche gain, which, in essence, is the average number of impact ionizations for a single carrier along the whole avalanche path. Both dark current and photocurrent are multiplied by avalanche, but be aware only the induced part of the current is focused when discussing avalanche gain. Avalanche gain is typically expressed as the following equation:

$$M = \frac{I - I_d}{I_1 - I_{d1}} = \frac{R}{R_1} \tag{18}$$

Where M is the avalanche gain, I and I_d are the total and dark current at a specified bias voltage, and I_1 and I_{d1} are the total and dark current at a referenced unit gain bias voltage where gain is 1 (unit gain point). Apparently, M can also be expressed in responsivity form where R and R_1 are responsivities at specified and referenced unit gain bias voltage, respectively. The selection of unit gain point is not constant, the overall rule is selecting a low bias voltage where R changes slowly as the bias voltage varying. M in this form thus could be extracted from I-V and R test and be plot in I-G-V curve form.

In the local avalanche theoretical model established by McIntyre in 1966, M at x could be deducted as the following formula [29]:

$$M(x) = \frac{exp\left[-\int_x^W (\alpha - \beta)dx\right]}{1 - \int_0^W \alpha \bullet exp\left[-\int_{x'}^W (\alpha - \beta)dx'\right]dx'} \tag{19}$$

The average $M(x)$ in multiplication region is the commonly used gain conception. The parameter W is the width of the multiplication region, x' is the place where any impact ionization on avalanche chain happens, x is the place where initial impact ionization happens. α and β are the impact ionization coefficients mentioned before, which play an important role in M. From the formula, it is obvious that increasing the thickness of the multiplication region will increase M, but higher noise could then be induced. Meanwhile decreasing k will benefit M. It is noteworthy that this formula only applies to the situation that the multiplication region is far longer than the dead space length of carriers, in which case the gain is only related to the local field, dead space length is the minimum accelerating length that is required for carriers before impact ionization occurs. For the case that the multiplication region size can be compared to the dead space length the gain will be related to the history path, and thus the dead space model of gain calculation should be adopted.

Figure 7 provides an example of the I-G-V curve of a mesa-structured Ge on Si APD reported before [30]. The punch-through and avalanche breakdown voltages were − 8 and − 27 V, respectively; for M calculation, the unit gain point was chosen as about −7.5 V where $R = 0.52\,A/W$. The punch-through and avalanche breakdown voltages could vary a lot according to different device structures and material systems. It is evident that the avalanche gain M increases as reverse bias voltage increases, and this is because higher reverse bias voltage causes a higher electric field in the multiplication area. As a result, the impact ionization coefficient is increased, and the avalanche gain M climbed. Moreover, it is important to declare punch-through voltage

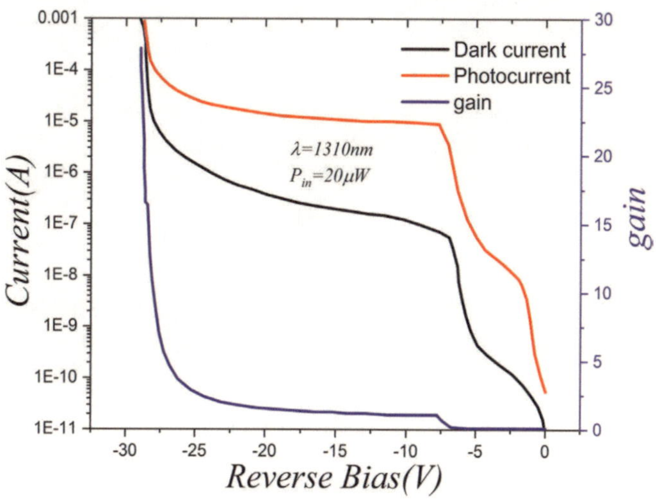

Figure 7.
I-G-V curve characteristics of a 30um-diameter Ge on Si APD at room temperature (Redraw from literature [30]).

and breakdown voltage. In general, punch-through means at a specified reverse bias that the whole device is fully depleted. Thus, photo-detection efficiency would be enhanced, breakdown voltage is defined as the bias voltage where avalanche is established, it could be confirmed by calculating the intercept of the linear fit of the avalanche voltage range of I-V result [31]. If an APD functions above breakdown voltage, it turns to a SPAD, and the avalanche cannot stop unless a quenching or gated mechanism is imposed.

The bandwidth of APD is decided by two factors: carrier transit time and RC constant. Carrier transit time is composed by carrier transport time which is decided by the carrier mobility and depletion region size and avalanche buildup time that is mainly decided by the gain and electric field in the multiplication region. The resistor R mainly comes from electrode contact, while the capacity C rises from the depletion region. The less RC value is, the higher bandwidth APD could achieve, which makes ohm contact of the electrodes essential, note that thicker absorption region could reduce capacity and improve responsivity indeed but the carrier transportation time also increases. There is a tradeoff between responsivity and bandwidth unless wave-guide structure is adopted, thicker structure could also introduce more noise. More-over, lower k could benefit the bandwidth [32].

Gain-bandwidth production (GBP) is a special standard of APD, the avalanche establish time will dominate the bandwidth property, with gain increasing the band-width decreasing since avalanche established time increases. As a result, GBP is close to a constant as gain exceeds 1/k as illustrated in **Figure 8** [32]. GBP is a critical standard of APDs, always being used to evaluate the performance of APDs for it is the most direct property related to communication system.

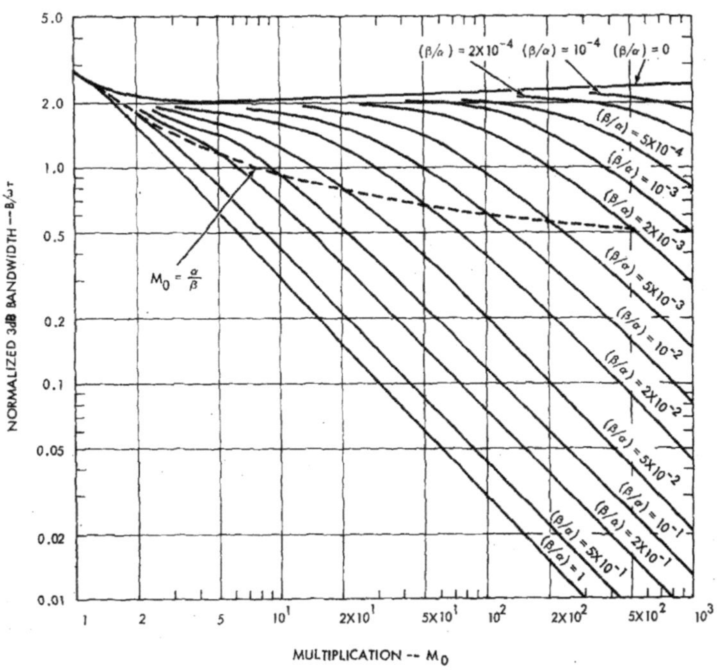

Figure 8.
Bandwidth as a function of multiplication in APDs [32] (Reprinted from [J. Appl. Phys. 38, 3705 (1967)], with the permission of AIP Publishing.) Copyright © 1967.

As discussed above, avalanche brings internal gain to APD. Therefore, higher responsivity is achieved, making APD an ideal candidate for weak signal detection. However, the avalanche could also induce noise for the randomness of impact ionizations, namely excess noise.

Besides the shot noise in typical PIN PDs discussed above, due to the presence of the excess noise, the noise in APDs S_{APD} together with the excess noise factor F could be expressed in the simplified form [29]:

$$S_{APD} = 2qIFM^2 \tag{20}$$

$$F = kM + (1-k)\left(2 - \frac{1}{M}\right) \tag{21}$$

The shot noise components q and I are the charge of the carrier and the total current, while the parameter F is the excess noise factor, which is strongly relevant to the impact ionization ratio k and the avalanche gain M as shown in **Figure 9**.

From **Figure 9**, it is straightforward that all curves start from $F = 1$ as $M = 1$ and slight change on k could bring dramatic lifting to the excess noise factor F of APD, especially F-M line shape will be linear ($F = M$) when k reaches 1 where F meets its maximum; on the contrary, if k keeps zero which is of course ideal F will asymptotic to 2 as M increasing. Reducing noise in APDs is an urgent task since the shot noise and the excess noise could increase the bit error rate (BER), thus affecting the sensitivity performance of receivers based on APDs.

Be aware that sensitivity is a generic key standard being used to describe the merit of a receiver system which is composed by detectors as PD/APD that convert the light signal into the current form; then, the current signal is converted to voltage form and further amplified by a series connected transimpedance amplifier (TIA) together with limiting amplifier (LA); in the end, the data are collected by CDR (clock and data recovery). The schematic diagram of an optical receiver is given in **Figure 10**.

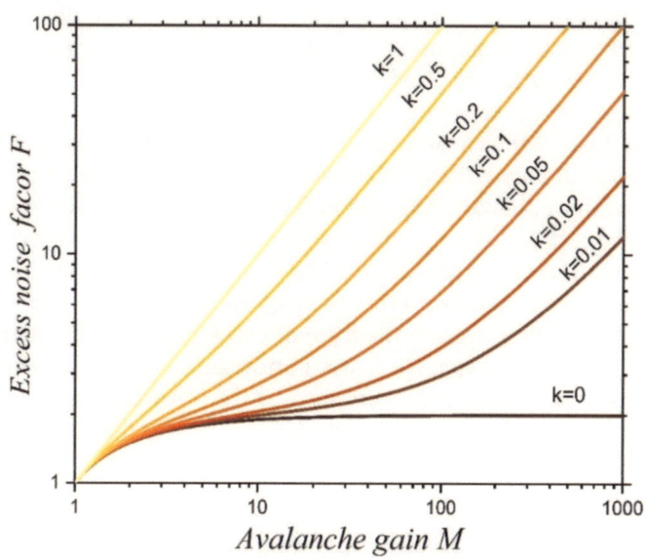

Figure 9.
The relationship among excess noise factor F, avalanche gain M, and impact ionization coefficient ratio k.

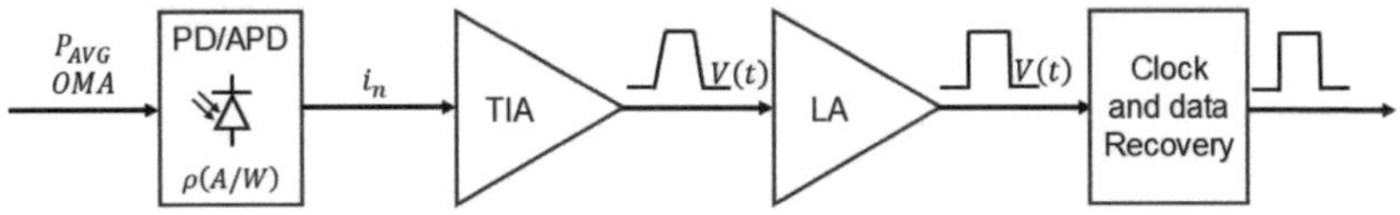

Figure 10.
The schematic diagram of an optical receiver. The parameters are explained in the content below.

The function of a TIA device could be analogically interpreted as a resistor, which transforms the current signal generated by PD/APD into voltage signal and further amplify the voltage signal. LA is used to set logic decision threshold, amplify and maintain the output voltage constant value. LA has a peak-to-peak operating signal swing range below which the output is not held constant. In an optical receiver, the noise mainly comes from PD/APD and amplifiers like TIA and LA. The major noise in an optical receiver comes from the detector part (PD/APD) and amplifier part (TIA, LA).

The receiver sensitivity is defined as the minimum incident optical power before BER exceeding some criteria (<10-12 for an example) and could be expressed in two forms, the average incident power (P_{AVG}) in dBm or the optical modulation amplitude (OMA) in W_{P-P} [33]:

$$P_{AVG} = 10\,log\left(\frac{i_n SNR(r_e+1)}{2R(r_e-1)}1000\right)\,dBm \tag{22}$$

$$OMA = \frac{i_n SNR}{R} \tag{23}$$

The prediction of receiver sensitivity thus could be realized adopting key parameters given by the vendors using the equation in which i_n is the input reference noise of TIA typically in A_{RMS}, i_n is directly proportional to the value of the photodiode capacitance and bandwidth of the TIA. SNR is the signal-to-noise ratio that is required to maintain a specific BER, and r_e is the extinction ratio which represents the ratio between logic 1 power level and a logic 0 power level; in addition, R is the responsivity of the PD/APD. From the equation, it is evident that the superior R induced by internal gain is the intrinsic origin of the better sensitivity of APD based receivers in comparison to the PIN PD counterparts.

It is noteworthy that the equation is oversimplified for the noise of PD/APD and LA are partly ignored; only noise of TIA is mainly considered while the PD/APD, TIA, and LA are major noise sources in receivers. Therefore, the equation is more suitable to compare the impact of different component on receiver sensitivity rather than directly obtaining the receiver sensitivity. The accurate sensitivity estimation relies on carefully designed BERT (bit error rate test) experiments, and the form of P_{AVG} is adopted more frequently instead of the form of OMA for the equipment measuring P_{AVG} is easier to acquire, the concept is straightforward as well. The sensitivity of single PD/APD device is usually characterized by the noise equivalent power NEP which represents the incident power when $SNR = 1$; in this discussion, the sensitivity of the receiver system is mainly focused since the performance of the whole system is more meaningful.

The receiver sensitivity could be affected by multiple factors, except for the noise generated in PD/APD, and amplifiers like TIA and LA as mentioned before, the modulation speed and format such as NRZ and PAM4 can also impose significant difference. Furthermore, receiver sensitivity could also change with different gain in

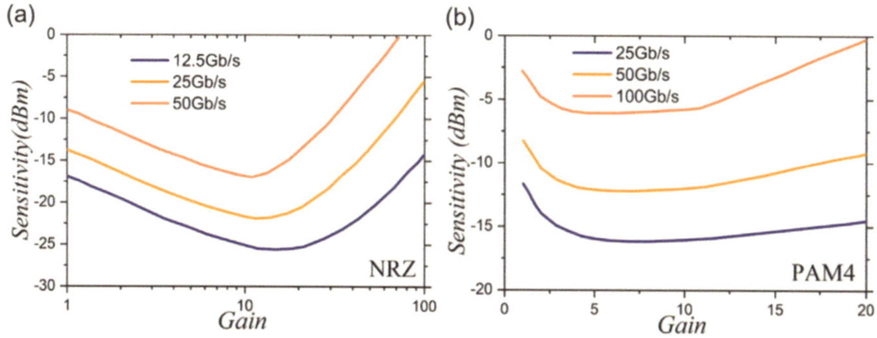

Figure 11.
The sensitivity of Ge on Si waveguide APD versus gain, data rate in data format of (a) NRZ and (b) PAM4 (Redraw from literature [34]).

APD receivers. Typically, receiver sensitivity decreases as data rate increasing or bits number in one single Bit, which means sensitivity of 10 Gb/s communication will be better than that of 25 Gb/s, and sensitivity of communication in NRZ format will be better than that in PAM4 format. In APD receiver, sensitivity usually first improves with increasing gain but then degenerates as gain comes to relatively high as shown in **Figure 11** [34].

There is a gain range starting from 1 inside which sensitivity increases with gain increases, and the APD receiver shows superior sensitivity in comparison to PIN PD receiver counterparts. Sensitivity will decrease above this gain range, and this range is also affected by the modulation speed and format. This phenomenon could be related to the excess noise discussed before, at low gain situation, sensitivity is improved due to the avalanche and the excess noise is controlled at a low level, as gain increasing the excess noise increases exponentially and eventually worsens the sensitivity. Therefore, attempts to eliminate the excess noise in APD is of great significance, otherwise, only low gain operation could be adopted in high-speed communication system.

4.3 Linear mode and Geiger mode operation

In brief, linear mode and Geiger mode for APD are divided by the breakdown voltage, SPAD could be considered as a kind of special APD that is biased above the breakdown voltage and thus obtains the ability to discern the smallest quanta of light, a single photon in another word. The structure of SPAD could be slightly modified compared to regular APD to optimize the performance. It is noteworthy that there are other types of single-photon detectors, such as superconducting nanowire single-photon detectors (SNSPD), which can only work at a cryogenic temperature of a few K [35], and the cryogenic working condition limits the application of SNSPD.

The most distinct feature of SPAD is that the avalanche cannot stop spontaneously, and a quenching circuit or self-quenching mechanism should be introduced to ensure the successful function of SPAD.

In a recently reported Ge on Si SPAD [36], the dark current is low to pA order. The merits of timing jitter, after-pulsing characteristic time, and after-pulsing probability in this report are, respectively, measured as $19\,\text{kHz}\,\mu\text{m}-2$, 12%, 188 ps, $\sim$90 ns, and $< 1\%$, with a low breakdown voltage of 10.26 V and a small excess bias of 0.75 V.

4.4 External factors affecting APD performance: Temperature, bias voltage, incident power, and wavelength

While APD performance is relevant to the intrinsic factors like the materials and the device structures indeed, it is important to declare that APD performance could also be affected by multiple factors such as operating temperature, bias voltage, incident power, wavelength, and even the cosmic ray when operating on outer space satellites and so on.

Temperature in semiconductors reflects on the thermal fluctuations, phonons, in other words. As the temperature climbs, the mean energy and number of phonons will increase. As a result, the electron-phonon coupling effect is enhanced, which limits the mobility of carriers, decreases the impact ionization rate, and further weakens the avalanche procedure.

The bias voltage will affect the avalanche by changing the electric field amplitude in the avalanche region, which is straightforward to imagine. Typically, gain increases with the applied bias voltage. There could be a reverse tendency that comes up at a higher bias voltage due to the negative differential resistance effect.

The incident power can also affect the gain performance of APD, which has been widely reported, and the regular trend is that gain increases with the decreasing incident power. This phenomenon is explained by the reduction of resistance at high gain [37].

The impact of different wavelengths on the performance of APD could be attributed to the wavelength-dependent absorption coefficient of material in the absorption region.

4.5 Applications of APDs: Optical fiber communication, LiDAR, and spectroscopy

It has never been an era in history like recent years that the technology iteration is so rapid. In recent years, emerging technics such as artificial intelligence (AI), internet of things (IoT), cloud computing and storage, and video streaming are thirsty for huge data communications. As a result, 5G communications, hyper-scale data centers, and exascale high-performance computers (HPC) have been developed [25]. To compensate for the power consumption demands of these newly developed techniques, APDs rise as proper candidates for the higher receiver sensitivity over the broad spectral wavelength range and high signal-to-noise ratios [38, 39]. APD in both linear mode and Geiger mode is widely utilized in communication [40], LiDAR [41], 3D imaging [42], quantum secret key transmission [43], spectroscopy [44], and single-photon detection [45]. Due to the quenching mechanism, SPAD operating in Geiger mode has a limited bandwidth of tens to hundreds of MHz, adapting to applications like LIDAR for lower NEP and higher sensitivity [36]. The LIDAR market faces fluctuations due to the serious challenge of the full-vision solution and the multiple potential technical paths such as dTOF, iTOF, AMCW, FMCW, and so many on [46]. Therefore, in the long term, light communication will still be the major battlefield of APDs. In the future, APD is also expected to improve the optical interconnection in data centers or server groups.

5. Challenges and limitations of photovoltaic detectors

5.1 Dark current and its reduction techniques

As mentioned in 3.2, the dark current of photodetector mainly originates from thermal noise, defects, and imperfections in crystals, grain boundaries, etc. Dark current has multiple impacts on detectors. It can affect key performance parameters of detectors such as signal-to-noise ratio, sensitivity, and power consumption.

Therefore, research on suppressing dark current technology has received widespread attention. One method is reducing impurities and defects in the device. Besides, the use of heterojunction structures [47], Schottky contacts, or two-dimensional material electrode contacts [48] can significantly reduce dark current. Surface leakage current can be improved by growing oxide layers, epoxy encapsulation, or depositing passivation layers [49]. In addition, adjusting device size and materials, such as adding SiO_2 layer in metal semiconductor metal (MSM) structure by Chuang et al., can increase the barrier height and reduce dark current [50]. What's more, using new device structures is a novel way to reduce dark current. For example, the dark current shunting (DCS) method proposed by Yang et al. can theoretically completely suppress dark current [51].

5.2 Noise sources and their impact on sensitivity

As discussed in the last section, noise in photovoltaic detectors is one of the major deterioration factors in optical receiver sensitivity. For PIN PD, the dominant noise source is the shot noise, while an extra multiplication of excess noise for APD is required because of the randomness of the avalanche multiplication. In detail, the randomness in the avalanche process is caused by the stochastic nature of the impact ionization, which eventually reflects on the avalanche gain fluctuations.

The shot noise comes from the Poisson distribution of the electron-hole pair generated by the incident photon stream, which is a stochastic process of random arriving times, in other words, with a given time interval, the incident photon number is certain but the number of photo-generated electron-hole pairs follows Poisson distribution, which means the responsivity could fluctuate with time due to the shot noise. Generally, DC photocurrent of I_{PD} will generate a shot noise power density of $\sigma_{sn}^2 = 2qI_{PD}$ according to the discussion before. Besides, there are other noise sources for an optical receiver system, such as ASE noise in optical amplifiers, which could be avoided in an APD receiver. All the noise mentioned here will degenerate the device sensitivity performance, thus further limiting the receiver sensitivity.

To reduce the noise in APD, several methods could be considered: (1) Choosing material system with natural low k. (2) Energy band engineering like impact ionization engineering (I^2E) method that could reduce effective k by suppressing the avalanche of unwanted carriers (holes when electrons are preferred to avalanche) [52, 53]. (3) Carefully design the thinner thickness of multiplication region using the non-local field theory to reduce the stochastic of ionization and further reduce excess noise [54]. (4) Superlattice multiplication region [55].

5.3 Responsivity and bandwidth trade-offs

There are several trade-offs in photovoltaic detectors, among which the trade-off between responsivity and bandwidth is one of the most important since both factors are vital. Responsivity and bandwidth trade-off are more prominent in normal incident vertical mesa structured PIN PD or APD in which the dimensions of the light incident and carrier transportation are the same; as the absorption or multiplication layer becomes thicker, the responsivity or gain is improved while the carrier transportation time is longer hence the bandwidth is limited. On the contrary, if the thickness is reduced, the bandwidth could be improved, yet the responsivity, together with the gain, could be decreased. Above, there is a tradeoff between responsivity and the bandwidth in mesa-structured photovoltaic detectors like PD or APD.

As mentioned before, responsivity-bandwidth trade-off mainly exists in PD/APD in which incident direction coincides with carrier transportation direction; the most representative structure is the normal incident vertical mesa structure. In order to overcome this annoying tradeoff, waveguide-structured PD/APD (WGPD/APD) is proposed in which the light incident direction is perpendicular to the carrier transportation direction. The light propagating in the waveguide is coupled into the absorption waveguide structure; the photon-generated carriers then drift along the direction perpendicular to that of the light incident, which means the responsivity-bandwidth tradeoff could be avoided.

It should be declared that there is no superiority between mesa PD/APD and WGPD/APD. Each of them has unique applications. Mesa devices make it easier to integrate 2D arrays for applications like space communication. WGPD/APD is more suitable for large-scale photonic integrated circuits (PICs) and on-chip optical interconnection. There are methods to further improve the responsivity of WGPD/APD, including further increasing the waveguide length, which is similar to improving the absorption region thickness of a normal-incident mesa PD/APD in principle, adding DBR or loop reflectors at the end of the waveguide [56, 57], or even sinking the absorption waveguide into the propagating waveguide [58].

5.4 Temperature dependence and stability

Inherently, photodetectors are sensitive to temperature. This is because many characteristics of photodetectors vary with temperature, such as carrier density, mobility, and generation rate. Thus, temperature control and compensation are essential for the stability of photodetectors.

The issue of temperature dependence is especially severe for APD. Avalanche gain and breakdown voltage are two main parameters that are highly affected by temperature. During the ionization impact or avalanche process, there exist a number of scattering mechanisms, which will reduce the energy of electrons (or holes) energy [59]. Among these scattering mechanisms, phonon scattering is the most sensitive to temperature and can influence the whole avalanche process. As temperature increases, the phonon scattering is strengthened, causing more energy loss of the avalanche carriers (electrons or holes). Thus, the carriers need to cross more distance in order to acquire the ionization threshold energy, resulting in a decrease in the ionization coefficient. In semiconductors such as silicon, GaAs, InP and AlGaAs, both the electron and hole ionization coefficient decrease as the temperature increases [60]. Therefore, the avalanche gain also decreases. Besides, due to stronger scattering, carriers also need a higher electric field to accelerate them to get enough kinetic energy in order to make impact ionization happen. So, the breakdown voltage also increases. Experiments and simulations have shown that, in InP/InGaAs and InP/AlGaAsSb APDs, the breakdown voltage increases linearly with temperature [59, 61].

6. Advanced photodetector technologies

In order to meet the rising demand for detectors with high quality and application systems, merits including high gain, low noise, large scale, high computing power, and low energy consumption are required. In this section, major advances in recent years will be discussed in three topics: linear high-gain APD, low-noise digital alloy APD, and large-scale high counting rate SPAD.

6.1 Linear high-gain APDs (GaN/AlN PSS APD)

Internal gain is the most prominent merit of APD, and higher linear gain is expected to achieve better sensitivity. One representative technical method is constructing periodically stacked structure APD (PSS APD), which is a kind of energy band engineering method. The schematic diagram comparing the working principles between traditional APD and PSS APD is given in **Figure 12** [62]. For example, AlN/GaN PSS APD with ultrahigh stable linear gain of 10,000 has been reported by experiment study [62]. AlN/GaN PSS model was proposed and carefully analyzed in order to explain the origin of the gain enhancement. Further, it is found that the lowest energy states for electrons are at the Γ valley in AlN, where the scattering rates are much lower than that in the ionization situation in the GaN bulk layer. As a result, the electrons that triggered interfacial ionization play an important role in the multiplication performance, which explains the ionization coefficient enhancement found by experiment [63]. Furthermore, GaN/AlN PSS APD with deep Γ valley and large conduction band offset was systematically investigated, and it has been found that decreasing scattering degree for electron transport is found to be necessary for realizing high-temperature stability and high ionization efficient APD [64]. Changing the orientation of the InAlAs digital alloy from [001] to either [110] or [111] has the potential to suppress the APD excess noise because the effective mass of holes in the valence band increases more rapidly than that of electrons in the conduction band, this has been proved by a theoretical analysis [65].

Both increasing periodical thickness and increasing AlN layer occupancy are preferred to achieve high gain under a low electric field, while the former benefits in decreasing the excess noise and the latter harms slightly in excess noise [66].

Besides, based on GaAsSb/AlGaAsSb, APD working at 1550 nm with high multiplication (gain>1200) and low excess noise ($F < 7$ at gain = 200) by optimizing the charge sheet has been fabricated [67].

Moreover, as is discussed before, in communication applications, the gain-bandwidth production is more often used as standard to estimate the performance of APD. Ge on Si APD possessing a large bandwidth of 53 GHz in the gain range of 9–19.5 and an ultrahigh GBP of 1033 GHz under −8.6 V and at 1550 nm is reported utilizing novel capacity and inductance structure modification [68].

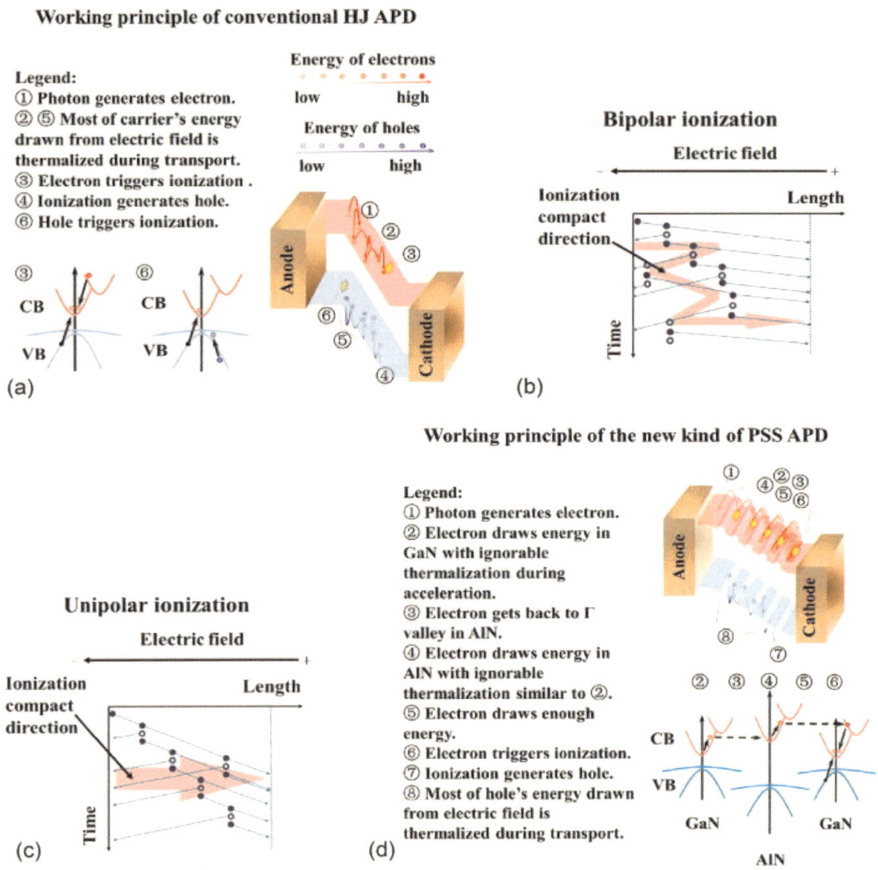

Figure 12.
(a) Carrier transport in conventional traditional APDs under electric field, (b) Bipolar ionization proceeding along the two opposite directions and (c) A unipolar ionization proceeding along a single direction. (d) Carrier transport in GaN/AlN PSS APDs [62] (Reprinted from [Applied Physics Letters 109, 241,105 (2016)], with the permission of AIP Publishing.) Copyright © 2016.

6.2 Digital alloy APD

Digital alloy is a relatively new topic in recent years that attract intense attention from a wide range of researchers, mainly refers to a special III-V material alloy system $Al_xIn_{1-x}As_ySb_{1-y}$ or $Al_xGa_{1-x}As_ySb_{1-y}$ that could serve as multiplication region in superlattice form [69]. The merits of digital alloys include low excess noise, high gain bandwidth, and tunable bandgap by changing the alloy composition or superlattice period thickness, which makes them strong competitors in the aspect of LIDAR systems and telecommunications.

The low noise merit of digital alloy APD originates from the low *k* value realized in the superlattice structure, in detail, the creation of minibands that localize carriers. For example, in a reported InAlAs digital alloy APD, the ionization coefficients of electrons in digital and random alloys are almost the same, but the ionization coefficient for holes in the digital alloy is significantly suppressed relative to the random alloy, which results in lower *k* value approaching to that of Si material [70].

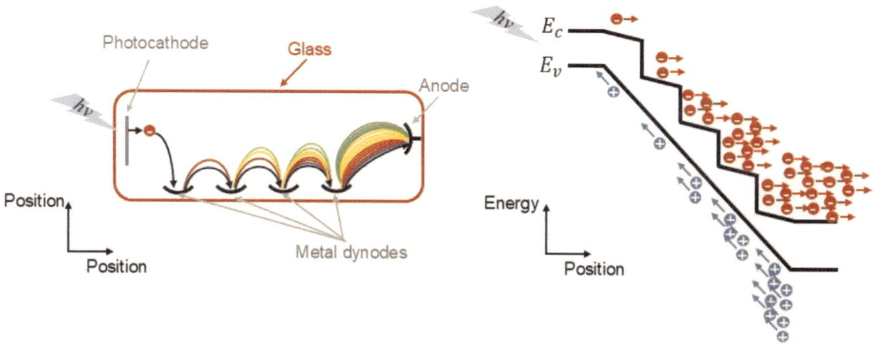

Figure 13.
(a) A PMT with four metallic dynode stages contained within a large glass vacuum. (b) A multistep staircase APD energy-position band diagram illustrating localized carrier multiplication at the step regions (Redrawn from literature [71]).

Utilizing the I^2E method discussed before, the multistep staircase APDs with extremely low noise and deterministic amplification were proposed adopting the digital alloys [71]. Different from typical APD that requires a high electric field to accelerate carriers to realize impact ionization, I^2E multistep staircase relies on the energy drop in energy band structure to accelerate carriers to initiate impact ionization as illustrated in the right part of **Figure 13**. This procedure could be interpreted by the analogy in PMT principles as shown in the left part of **Figure 13**. Typical PMTs have between 6 to 13 dynodes, all separated by >100 V potentials. Primary photoelectrons generate a cascade of high-energy secondary electrons that acquire energy from the large electric fields between adjacent dynodes. The increase in the secondary electron count over the primary electron count is registered as a gain in the form of photocurrent at the anode. The right part of **Figure 13** shows that APD electron-initiated impact ionization produces secondary electrons in the conduction band (Ec) and secondary holes in the valence band (Ev). The lack of valence band discontinuity prevents hole-initiated impact ionization. Localized, single-carrier-initiated impact ionization events lead to low-noise performance.

Moreover, according to a newly reported result, $In_{0.53}Ga_{0.47}As/Al_{0.85}Ga_{0.15}As_{0.56}Sb_{0.44}$ based APD has achieved extremely low excess noise factor of less than 3.5 at gain of 100, an extremely low NEP of 21.2 fW/$\sqrt{\text{Hz}}$ was achieved with a 30-μm-diameter APD, over a 440 MHz signal bandwidth, at an APD gain of 230 and the wavelength of 1550 nm. Current commercial APD-TIA modules typically exhibit NEPs of >100 fW/$\sqrt{\text{Hz}}$ to 10's of pW/$\sqrt{\text{Hz}}$. Compared to its best in-class 80-μm counterpart, this work's receiver demonstrated a 6.5× sensitivity improvement at 2× the operating bandwidth, with an NEP of 32.5 fW/$\sqrt{\text{Hz}}$ [72]. Furthermore, InAlAs could also be fabricated in digital alloy method instead of in traditional random alloy way, and digital alloy InAlAs has shown gain-bandwidth production of more than 140GHz [73].

6.3 Single-photon avalanche diode (SPAD dynamic memristor quenching)

There are several ways to improve the performance of SPAD, among which a major breakthrough is the dynamic memristor (DM), which could be planted in SPAD system circuits and functions as quenching circuits without extra space consuming

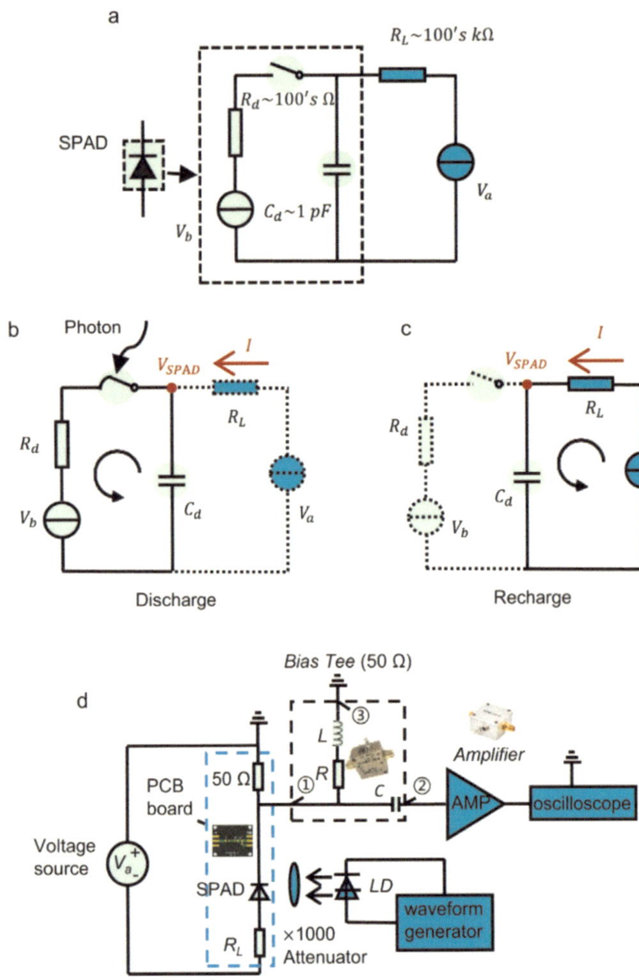

Figure 14.
(a) Equivalent circuit model for passive quenching of the SPAD in series with a quenching resistor. The dashed box is the SPAD, represented by a capacitor, a resistor, a photon switch, and a voltage source. (b) The SPAD discharge process when an avalanche is triggered by an incident photon. (c) Recharging process after quenching. (d) Single-photon detection system. Passive quenching is accomplished by connecting a SPAD and a resistor in series. A voltage source (Keithley 2400) is used to reverse bias the SPAD. The single-photon signal is generated by attenuating the optical signal from a laser diode driven by a Keysight 33600A waveform generator. The avalanche response current is determined from the voltage on a 50 Ω readout resistor, and the voltage is introduced to an amplifier (Mini-circuits ZFL-1000LN+) and then into an oscilloscope (Rigol DS7024) through a bias tee (Mini-circuits ZFBT-4R2GW+) [74]. CC BY 4.0.

and RC delay sacrificing brought by traditional resistive quenching. DM is a device whose resistance can be modulated by voltage. At the moment of an avalanche, the DM is in a high resistance state (HRS), and it quickly divides the voltage, causing the avalanche to stop. Subsequently, the voltage division on the DM decreases its resistance, the RC time constant is rapidly reduced, and the avalanche detector quickly charges to recover the voltage. The equivalent circuits and the system diagram are given in **Figure 14** for Ref. [74]. Moreover, it has been proved that DM is faster in quenching the SPAD in a large size SPAD quenched by dynamic memristor, especially when the photon signal repetition rate is over 1 MHz [75, 76].

7. Conclusion

7.1 Summary of key points

In summary, the development history of photodetectors from three aspects: wavelength, material, and structure, and three common types of photodetectors are introduced: PIN PD, APD, and SPADs. By analyzing the noise sources and origin of dark current, the relationship between the device's sensitivity performance and its material and device characteristics is bridged. From these analyses, the readers could understand that the sensitivity of PIN PD, especially under high data rates, can be significantly enhanced by merely increasing the bias voltage to turn it into APD mode. A step further, this chapter provides deeper insights into modifying the APD structure from the perspective of carrier transportation and impact ionization engineering. Besides, Geiger mode APD, which is capable of sensing single photon level weak light signal, is also described. In the final part of the chapter, the challenges of these detectors are discussed, and several recent technical approaches, including linear high-gain APD, digital alloys APD, and SPAD with dynamic memristor quenching, have been presented to tackle these challenges. By demonstrating the basic principles of photovoltaic detectors, particularly APD, to the analysis of receiver sensitivity, this chapter could provide comprehensive and valuable insights for the design and optimization of high-sensitivity detection systems. These systems find applications in various fields, including optical interconnection, optical communication, LiDAR, and so on.

7.2 Future trends and challenges in photodetector technology

Photodetectors will develop towards higher sensitivity, faster response speed, wider wave band, and lower power consumption. For example, the innovation of single-photon detectors and APD technology have significantly improved detection efficiency and response capability [68, 77].

New materials such as two-dimensional materials [78], perovskites [79], quantum dots [80], etc., have become essential research directions for photodetectors due to their superior optical properties such as higher absorption coefficients, higher carrier mobility, and higher crystal quality, which help improve the sensitivity and response speed of detectors.

Photodetectors are developing towards integration and miniaturization to meet the needs of intelligent devices. For example, breakthroughs have been made in integrating and miniaturizing organic material photodetectors [81] and two-dimensional material detectors.

Photodetectors are also becoming more intelligent and multifunctional, integrating multiple functions such as spectral detection, polarization detection, etc. For example, single-photon LiDAR technology achieves high-precision 3D reconstruction through time-dependent single-photon technology [82]. Single photon detectors have essential application prospects in weak light detection, quantum communication, and other fields due to their high sensitivity and low noise characteristics [83].

Research on ultrawideband photodetectors can meet the needs of broadband detection, especially in fields such as remote sensing and medical imaging.

The future development of photodetectors also faces several challenges. There are some technical bottlenecks remain to be solved, such as responsivity-bandwidth tradeoff, reliability, dark current etc.

Inherently, photodetectors face the contradiction between responsivity and bandwidth, especially for ultrawideband photodetectors. Reliability under high temperatures is another major challenge faced by photodetectors [84], especially in the era of artificial intelligence.

To reduce dark current and improve sensitivity, it is necessary to study the internal physical mechanism of the detector, as well as the preparation process and device structure.

7.3 Potential applications and societal impact

In the era of artificial intelligence and the Internet of Things, the arrival of AGI has put forward higher requirements for communication and perception. The application of photodetectors in multiple fields, such as communication, transportation, healthcare, and city management, continues to deepen. With the popularization of 5G and IoT technologies, the demand for photodetectors in high-speed data transmission and remote monitoring will continue to grow, driving the development of the communication industry [85]. In addition, applying photodetectors in vehicle-mounted LiDAR will promote the development of autonomous driving technology and enhance the safety and efficiency of intelligent transportation systems [86]. Meanwhile, applying photodetectors in biomedical imaging, gene sequencing, optical coherence tomography (OCT), and other fields will significantly improve the accuracy and efficiency of medical diagnosis, promoting the development of precision medicine [87]. Furthermore, photodetectors are widely used in smart city construction for intelligent transportation, environmental monitoring, and energy management, enhancing urban intelligence [88]. In a word, photodetectors play essential roles in numerous fields of our lives and have become crucial building blocks of information society.

Author details

Jiyuan Zheng[1*†], Xiaoran Wang[2†], Beichen Liu[2†], Rui Wu[2†], Jizhe Zhao[2], Yinjie Liu[2], Xiayang Hua[2], Yubo Yang[1,2], Zhibiao Hao[1,2], Anran Guo[3] and Lai Wang[1,2*]

1 Beijing National Research Center for Information Science and Technology, Tsinghua University, Beijing, China

2 Department of Electronic Engineering, Tsinghua University, Beijing, China

3 Department of Solid State Image Sensor, CETC No.44 Research Institute, CETC Chips Technology Group CO., LTD, Chongqing, P.R. China

*Address all correspondence to: zhengjiyuan@tsinghua.edu.cn and wanglai@tsinghua.edu.cn

† These authors contributed equally.

References

[1] Adams WG, Day RE. V. The action of light on selenium. Proceedings of the Royal Society of London. 1877;**25** (171-178):113-117. DOI: 10.1098/rspl.1876.0024

[2] Sawyer DE, Rediker RH. Narrow base germanium photodiodes. Proceedings of the IRE. 1958;**46**(6):1122-1130. DOI: 10.1109/JRPROC.1958.286894

[3] Lucovsky G, Cholet P. GaAs, a sensitive photodiode for the visible. Journal of the Optical Society of America. 1960;**50**(10):979-983. DOI: 10.1364/JOSA.50.000979

[4] Chapin DM, Fuller CS, Pearson GL. A new silicon p-n junction photocell for converting solar radiation into electrical power. Journal of Applied Physics. 1954; **25**(5):676. DOI: 10.1063/1.1721711

[5] Lasser M, Cholet P, Wurst E Jr. High-sensitivity crystal infrared detectors. Journal of the Optical Society of America. 1958;**48**(7):468-473. DOI: 10.1364/JOSA.48.000468

[6] Cao F, Liu Y, Liu M, Han Z, Xu X, Fan Q, et al. Wide bandgap semiconductors for ultraviolet photodetectors: Approaches, applications, and prospects. Research. 2024;7:0385. DOI: 10.34133/research.0385

[7] Webb P, McIntyre R, Holunga M, et al. Planar InGaAs/InP APD fabrication using silicon implantation and regrowth techniques. Components for Fiber Optic Applications II. 1988;**839**:148-154. DOI: 10.1117/12.942559

[8] Zhang Y-G, Gu Y, Shao X-M, Li X, Gong H-M, Fang J-X. Short-wave infrared InGaAs photodetectors and focal plane arrays. Chinese Physics B. 2018;**27**(12):128102. DOI: 10.1088/1674-1056/27/12/128102

[9] Jansen H, de Boer M, Legtenberg R, Elwenspoek M. The black silicon method: A universal method for determining the parameter setting of a fluorine-based reactive ion etcher in deep silicon trench etching with profile control. Journal of Micromechanics and Microengineering. 1995;**5**(2):115. DOI: 10.1088/0960-1317/5/2/015

[10] Sarkar S, Elsayed AA, Sabry YM, Marty F, Drévillon J, Liu X, et al. Black silicon revisited as an ultrabroadband perfect infrared absorber over 20 μm wavelength range. Advanced Photonics Research. 2022;**4**(2):2200223. DOI: 10.1002/adpr.202200223

[11] Dawi EA, Karar AA, Mustafa E, Nur O. Plasmon-enhanced light absorption in (p-i-n) junction GaAs nanowire solar cells: An FDTD simulation method study. Nanoscale Research Letters. 2021;**16**(1):149. DOI: 10.1186/s11671-021-03603-1

[12] Zeng LH, Wang MZ, Hu H, Nie B, Yu YQ, Wu CY, et al. Monolayer graphene/germanium Schottky junction as high-performance self-driven infrared light photodetector. ACS Applied Materials & Interfaces. 2013;**5**(19):9362-9366. DOI: 10.1021/am4026505

[13] Shen J, Liu X, Song X, Li X, Wang J, Zhou Q, et al. High-performance Schottky heterojunction photodetector with directly grown graphene nanowalls as electrodes. Nanoscale. 2017;**9**(18):6020-6025. DOI: 10.1039/c7nr00573c

[14] Hu C. Modern Semiconductor Devices for Integrated Circuits. Upper Saddle River, N.J.: Prentice Hall; 2010

[15] Anderson B, Anderson R. Fundamentals of Semiconductor Devices. New York, NY: McGraw-Hill, Inc.; 2004

[16] Pierret RF. Semiconductor Device Fundamentals. Boston, MA: Addison-Wesley Publishing Company; 1996

[17] Neamen DA. Semiconductor Physics and Devices: Basic Principles. New York, NY: McGraw-Hill; 2003

[18] Liu E, Zhu B, Luo J. The Physics of Semiconductors. Beijing: Publishing House of Electronics Industry; 2008

[19] Chen Z, Lei T, Ma J. A Concise Course in Semiconductor. Physics: China Machine Press; 2011

[20] Xie C, Mak C, Tao X, Yan F. Photodetectors based on two-dimensional layered materials beyond graphene. Advanced Functional Materials. 2017;**27**(19):1603886. DOI: 10.1002/adfm.201603886

[21] Yao Y, Zhang L, Yu J, Gao J, Huang C, Liu Y, et al. Gate-voltage control of waveguide-coupled lateral Ge-on-Si APD for 1550 nm detection. IEEE Sensors Journal. 2024;**24**(23):38929-38934. DOI: 10.1109/JSEN.2024.3481461

[22] Zimmermann H. Ultra-Sensitive PIN and Avalanche Photodiode Receivers. Philadelphia, PA: IOP Publishing; 2023

[23] Keshner MS. 1/f noise. Proceedings of the IEEE. 1982;**70**(3):212-218. DOI: 10.1109/PROC.1982.12282

[24] Campbell JC, Demiguel S, Ma F, Beck A, Guo X, Wang S, et al. Recent advances in avalanche photodiodes. IEEE Journal of Selected Topics in Quantum Electronics. 2004;**10**(4):777-787. DOI: 10.1109/jstqe.2004.833971

[25] Wang B, Mu J. High-speed Si-Ge avalanche photodiodes. PhotoniX. 2022;**3**:1. DOI: 10.1186/s43074-022-00052-6

[26] Tarof L. Planar InP-InGaAs avalanche photodetectors with n-multiplication layer exhibiting a very high gain-bandwidth product. IEEE Photonics Technology Letters. 1990;**2**(9):643-646. DOI: 10.1109/68.59337

[27] Campbell JC, Dentai A, Holden W, Kasper B. High-performance avalanche photodiode with separate absorption 'grading' and multiplication regions. Electronics Letters. 1983;**19**(20):818-820. DOI: 10.1049/el:19830558

[28] Lenox C, Nie H, Yuan P, Kinsey G, Homles A, Streetman B, et al. Resonant-cavity InGaAs-InAlAs avalanche photodiodes with gain-bandwidth product of 290 GHz. IEEE Photonics Technology Letters. 1999;**11**(9):1162-1164. DOI: 10.1109/68.784238

[29] McIntyre R. Multiplication noise in uniform avalanche diodes. IEEE Transactions on Electron Devices. 1966;**1**:164-168. DOI: 10.1109/T-ED.1966.15651

[30] Kang Y, Litski S, Sarid G, Morse M, Paniccia MJ, Pauchard A, et al. Ge/Si avalanche photodiodes for 1.3 μm optical fiber links. In: 2007 4th IEEE International Conference on Group IV Photonics. Piscataway, N.J., IEEE; 2007

[31] Zheng J-Y, Wang L, Hao Z-B, Luo Y, Wang L-X, Chen X-K. A GaN p—i—p—i—n ultraviolet avalanche photodiode. Chinese Physics Letters. 2012;**29**(9):097804. DOI: 10.1088/0256-307X/29/9/097804

[32] Emmons R. Avalanche-photodiode frequency response. Journal of Applied Physics. 1967;**38**(9):3705-3714. DOI: 10.1063/1.1710199

[33] Angelov K, Sadinov S, Varbanova N. Optical receiver sensitivity evaluation in presence of noise in digital communication system. In: ICEST 2012. Vol. 1. Piscataway, N.J.: IEEE; 2012. pp. 85-88

[34] Wang B, Huang Z, Sorin WV, Zeng X, Liang D, Fiorentino M, et al. A low-voltage Si-Ge avalanche photodiode for high-speed and energy efficient silicon photonic links. Journal of Lightwave Technology. 2019;**38**(12): 3156-3163. DOI: 10.1109/JLT.2019.2963292

[35] Najafi F, Mower J, Harris NC, Bellei F, Dane A, Lee C, et al. On-chip detection of non-classical light by scalable integration of single-photon detectors. Nature Communications. 2015;**6**(1):5873. DOI: 10.1038/ncomms6873

[36] Na N, Lu Y-C, Liu Y-H, Chen P-W, Lai Y-C, Lin Y-R, et al. Room temperature operation of germanium–silicon single-photon avalanche diode. Nature. 2024;**627**(8003):295-300. DOI: 10.1038/s41586-024-07076-x

[37] Piels M, Ramaswamy A, Zaoui WS, Bowers JE, Kang Y, Morse M, et al. Microwave nonlinearities in Ge/Si avalanche photodiodes having a gain-bandwidth product of 300 GHz. In: Optical Fiber Communication Conference. Washington DC: Optica Publishing Group; 2009

[38] Campbell JC. Recent advances in telecommunications avalanche photodiodes. Journal of Lightwave Technology. 2007;**25**(1):109-121. DOI: 10.1109/jlt.2006.888481

[39] Nada M, Nakajima F, Yoshimatsu T, Nakanishi Y, Tatsumi S, Yamada Y, et al. High-speed III-V based avalanche photodiodes for optical communications—The forefront and expanding applications. Applied Physics Letters. 2020;**116**(14):140502. DOI: 10.1063/5.0003573

[40] Srinivasan SA, Berciano M, De Heyn P, Lardenois S, Pantouvaki M, Van Campenhout J. 27 GHz silicon-contacted waveguide-coupled Ge/Si avalanche photodiode. Journal of Lightwave Technology. 2020;**38**(11):3044-3050

[41] Albota MA, Heinrichs RM, Kocher DG, Fouche DG, Player BE, O'Brien ME, et al. Three-dimensional imaging laser radar with a photon-counting avalanche photodiode array and microchip laser. Applied Optics. 2002;**41**(36):7671-7678. DOI: 10.1364/AO.41.007671

[42] Niclass C, Rochas A, Besse P-A, Charbon E. Design and characterization of a CMOS 3-D image sensor based on single photon avalanche diodes. IEEE Journal of Solid-State Circuits. 2005; **40**(9):1847-1854. DOI: 10.1109/jssc.2005.848173

[43] Lydersen L, Wiechers C, Wittmann C, Elser D, Skaar J, Makarov V. Hacking commercial quantum cryptography systems by tailored bright illumination. Nature Photonics. 2010;**4**(10):686-689. DOI: 10.1038/nphoton.2010.214

[44] McClish M, Farrell R, Myers R, Olschner F, Entine G, Shah K. Recent advances of planar silicon APD technology. Nuclear Instruments and Methods in Physics Research Section A: Accelerators, Spectrometers, Detectors and Associated Equipment. 2006;**567**(1): 36-40. DOI: 10.1016/j.nima.2006.05.055

[45] Hofbauer M, Steindl B, Zimmermann H. Temperature dependence of dark count rate and after pulsing of a single-photon avalanche

diode with an integrated active quenching circuit in 0.35 μm CMOS. Journal of Sensors. 2018;**2018**(1): 9585931. DOI: 10.1155/2018/9585931

[46] Li N, Ho CP, Xue J, Lim LW, Chen G, Fu YH, et al. A progress review on solid-state LiDAR and nanophotonics-based LiDAR sensors. Laser & Photonics Reviews. 2022;**16**(11): 2100511. DOI: 10.1002/lpor.202100511

[47] Yang M, Yang Y, Liu Q, Zhou H, Han J, Xie X, et al. A 3D topological Dirac semimetal/MoO 3 thin film heterojunction infrared photodetector with a current reversal phenomenon. Journal of Materials Chemistry C. 2020; **8**(45):16024-16031. DOI: 10.1039/d0tc03374j

[48] Long M, Wang P, Fang H, Hu W. Progress, challenges, and opportunities for 2D material based photodetectors. Advanced Functional Materials. 2019; **29**(19):1803807. DOI: 10.1002/adfm.201803807

[49] Dong Y, Wang W, Lei D, Gong X, Zhou Q, Lee SY, et al. Suppression of dark current in germanium-tin on silicon pin photodiode by a silicon surface passivation technique. Optics Express. 2015;**23**(14):18611-18619. DOI: 10.1364/OE.23.018611

[50] Chuang RW, Huang Y-H, Tsai T-H. Germanium-tin (GeSn) metal-semiconductor-metal (MSM) near-infrared photodetectors. Micromachines. 2022;**13**(10):1733. DOI: 10.3390/mi13101733

[51] Jin P, Tang Y, Li D, Wang Y, Ran P, Zhou C, et al. Realizing nearly-zero dark current and ultrahigh signal-to-noise ratio perovskite X-ray detector and image array by dark-current-shunting strategy. Nature Communications. 2023; **14**(1):626. DOI: 10.1038/s41467-023-36313-6

[52] Ren M, Maddox S, Chen Y, Woodson M, Campbell JC, Bank S. AlInAsSb/GaSb staircase avalanche photodiode. Applied Physics Letters. 2016;**108**(8):081101. DOI: 10.1063/1.4942370

[53] Guo B, Schwartz M, Kodati SH, McNicholas KM, Jung H, Lee S, et al. InGaAs/AlInAsSb avalanche photodiodes with low noise and strong temperature stability. APL Photonics. 2023;**8**(11):116112. DOI: 10.1063/5.0168134

[54] Hayat MM, Sargeant WL, Saleh BE. Effect of dead space on gain and noise in Si and GaAs avalanche photodiodes. IEEE Journal of Quantum Electronics. 1992;**28**(5):1360-1365. DOI: 10.1109/3.135278

[55] Dehzangi A, Li J, Gautam L, Razeghi M. Avalanche photodetector based on InAs/InSb superlattice. Quantum Reports. 2020;**2**(4):591-599. DOI: 10.3390/quantum2040041

[56] Wang B, Huang Z, Yuan Y, Liang D, Zeng X, Fiorentino M, et al. 64 Gb/s low-voltage waveguide SiGe avalanche photodiodes with distributed Bragg reflectors. Photonics Research. 2020;**8**(7):1118-1123. DOI: 10.1364/PRJ.390339

[57] Yuan Y, Huang Z, Zeng X, Liang D, Sorin WV, Fiorentino M, et al. High responsivity Si-Ge waveguide avalanche photodiodes enhanced by loop reflector. IEEE Journal of Selected Topics in Quantum Electronics. 2021;**28**(2, Optical Detectors):1-8. DOI: 10.1109/JSTQE.2021.3087416

[58] Hu Y, Liu Z, Zhu Y, Liu Z, Song Z, Huang Q, et al. High-speed waveguide lateral Ge/Si avalanche photodetector for

C-band and L-band. Optics Express. 2024;**32**(15):25598-25606. DOI: 10.1364/OE.522133

[59] Abdullah S, Tan CH, Zhou X, Zhang S, Pinel L, Ng JS. Investigation of temperature and temporal stability of AlGaAsSb avalanche photodiodes. Optics Express. 2017;**25**(26):33610-33616. DOI: 10.1364/OE.25.033610

[60] Jin X, Xie S, Liang B, Yi X, Lewis H, Lim LW, et al. Temperature dependence of the impact ionization coefficients in AlAsSb lattice matched to InP. IEEE Journal of Selected Topics in Quantum Electronics. 2021;**28**(2: Optical Detectors):1-8. DOI: 10.1109/JSTQE.2021.3099912

[61] Xiao Y, Deen M. Temperature dependent studies of InP/InGaAs avalanche photodiodes based on time domain modeling. IEEE Transactions on Electron Devices. 2001;**48**(4):661-670. DOI: 10.1109/16.915678

[62] Zheng J, Wang L, Wu X, Hao Z, Sun C, Xiong B, et al. A PMT-like high gain avalanche photodiode based on GaN/AlN periodically stacked structure. Applied Physics Letters. 2016;**109**(24):241105. DOI: 10.1063/1.4972397

[63] Zheng J, Wang L, Wu X, Hao Z, Sun C, Xiong B, et al. Theoretical study on interfacial impact ionization in AlN/GaN periodically stacked structure. Applied Physics Express. 2017;**10**(7):071002. DOI: 10.7567/apex.10.071002

[64] Zheng J, Wang L, Yang D, Yu J, Meng X, Hao Z, et al. Low-temperature-dependent property in an avalanche photodiode based on GaN/AlN periodically-stacked structure. Scientific Reports. 2016;**6**(1):35978. DOI: 10.1038/srep35978

[65] Zheng J, Tan Y, Yuan Y, Ghosh AW, Campbell JC. Tuning of energy dispersion properties in InAlAs digital alloys. Journal of Applied Physics. 2019;**125**(24):245702. DOI: 10.1063/1.5091694

[66] Zheng J, Wang L, Wu X, Hao Z, Sun C, Xiong B, et al. The influence of structure parameter on GaN/AlN periodically stacked structure avalanche photodiode. IEEE Photonics Technology Letters. 2017;**29**(24):2187-2190. DOI: 10.1109/lpt.2017.2766454

[67] Liu Y, Jin X, Lee S, Jung H, Lewis H, Guo B, et al. Very high gain and low noise GaAsSb/AlGaAsSb avalanche photodiodes for 1550nm detection at room temperature. Optical Components and Materials XXI. 2024;**12882**:128820O

[68] Shi Y, Li X, Chen G, Zou M, Cai H, Yu Y, et al. Avalanche photodiode with ultrahigh gain–bandwidth product of 1,033GHz. Nature Photonics. 2024;**18**(6):610-616. DOI: 10.1038/s41566-024-01421-2

[69] Guo Y, Yu J, Chen W, Zhu L, Lu L. Multi-step multiplication structure with low excess noise for the AlInAsSb avalanche photodiode. Journal of Electronic Materials. 2024;**53**(12):8078-8086. DOI: 10.1007/s11664-024-11436-8

[70] Zheng J, Yuan Y, Tan Y, Peng Y, Rockwell AK, Bank SR, et al. Digital alloy InAlAs avalanche photodiodes. Journal of Lightwave Technology. 2018;**36**(17):3580-3585. DOI: 10.1109/jlt.2018.2844114

[71] March SD, Jones AH, Campbell JC, Bank SR. Multistep staircase avalanche photodiodes with extremely low noise and deterministic amplification. Nature Photonics. 2021;**15**(6):468-474. DOI: 10.1038/s41566-021-00814-x

[72] Sheridan B, Collins X, Taylor-Mew J, White B, Ng JS, Tan CH. An extremely low noise-equivalent power photoreceiver using high-gain InGaAs/AlGaAsSb APDs. Journal of Lightwave Technology. 2024;**43**(2):741-746. DOI: 10.1109/jlt.2024.3447284

[73] Wang W, Yao J, Li L, Ge H, Wang L, Zhu L, et al. High-speed InAlAs digital alloy avalanche photodiode. Applied Physics Letters. 2023;**123**(19):191102. DOI: 10.1063/5.0169935

[74] Zheng J, Xue X, Ji C, Yuan Y, Sun K, Rosenmann D, et al. Dynamic-quenching of a single-photon avalanche photodetector using an adaptive resistive switch. Nature Communications. 2022; **13**(1):1517. DOI: 10.1038/s41467-022-29195-7

[75] Zheng J, Zhao J, Yang Y, Liu Y, Hua X, Wang L, et al. Large-size Geiger-mode avalanche photodiode quenched by dynamic Memristor. In: 2024 IEEE Photonics Conference (IPC). Piscataway, N.J., IEEE; 2024. pp. 1-2

[76] Zhao J, Zheng J, Yang Y, Liu Y, Hua X, Wang L, et al. Enhancing large-area Geiger-mode avalanche photodiode performance through dynamic memristor quenching: A study on improving count rate, reducing jitter and mitigating afterpulsing. Applied Physics Express. 2024;**17**(7):074501. DOI: 10.35848/1882-0786/ad5512

[77] Hao H, Zhao Q-Y, Huang Y-H, Deng J, Yang F, Ru S-Y, et al. A compact multi-pixel superconducting nanowire single-photon detector array supporting gigabit space-to-ground communications. Light: Science & Applications. 2024;**13**(1):25. DOI: 10.1038/s41377-023-01374-1

[78] Wang Y, Mei L, Li Y, Xia X, Cui N, Long G, et al. Integration of two-dimensional materials based photodetectors for on-chip applications. Physics Reports. 2024;**1081**:1-46. DOI: 10.1016/j.physrep.2024.06.001

[79] Rogalski A. A look at the future of perovskite detectors. Applied Physics Letters. 2024;**125**(9):090501. DOI: 10.1063/5.0228001

[80] Zou T, Choi T, Liu A, Zhu H, Noh Y-Y. Printed quantum dot photodetectors for applications from the high-energy to the infrared region. Nano Energy. 2024;**125**:109539. DOI: 10.1016/j.nanoen.2024.109539

[81] He X, Li Y, Yu H, Zhou G, Ke L, Yip H-L, et al. A microsized optical spectrometer based on an organic photodetector with an electrically tunable spectral response. Nature Electronics. 2024;**7**(8):694-704. DOI: 10.1038/s41928-024-01199-9

[82] Staffas T, Elshaari A, Zwiller V. Frequency modulated continuous wave and time of flight LIDAR with single photons: A comparison. Optics Express. 2024;**32**(5):7332-7341. DOI: 10.1364/OE.508004

[83] Sullivan NM, Braverman B, Upham J, Boyd RW. Photon number resolving detection with a single-photon detector and adaptive storage loop. New Journal of Physics. 2024;**26**(4):043026. DOI: 10.1088/1367-2630/ad3be5

[84] Chen Y, Li Y, Niu S, Yang X, Dou W, Shan C, et al. High temperature resistant solar-blind ultraviolet photosensor for neuromorphic computing and cryptography. Advanced Functional Materials. 2024;**34**(24): 2315383. DOI: 10.1002/adfm.202315383

[85] Zhu Y, Chen H, Han R, Qin H, Yao Z, Liu H, et al. High-speed flexible

near-infrared organic photodiode for optical communication. National Science Review. 2024;**11**(3):nwad311. DOI: 10.1093/nsr/nwad311

[86] Min L, Sun H, Guo L, Zhou Y, Wang M, Cao F, et al. Pyroelectric-accelerated perovskite photodetector for picosecond light detection and ranging. Advanced Materials. 2024;**36**(26):2400279. DOI: 10.1002/adma.202400279

[87] Saleem MI, Kyaw AKK, Hur J. Infrared photodetectors: Recent advances and challenges toward innovation for image sensing applications. Advanced Optical Materials. 2024;**12**(33):2401625. DOI: 10.1002/adom.202401625

[88] Wang J, Ling C, Xue X, Ji H, Rong C, Xue Q, et al. Self-powered and broadband photodetectors based on high-performance mixed dimensional Sb2O3/PdTe2/Si heterojunction for multiplex environmental monitoring. Small. 2024;**20**(23):2310107. DOI: 10.1002/smll.202310107

Chapter 3

2D Materials and their Application to Near Infrared Detection

Vinoth Elangovan, Naskath Sithara H,
Harish Santhana Krishnan, Navaneethan Mani
and Jayaram Archana

Abstract

Due to their exceptional performance across a wide range of wavelength regions, from UV to broadband detection, 2D materials (2DMs)-based photodetectors have expanded quickly in recent years. 2DMs gained a reputation for playing important roles in photodetection because of their unique physical characteristics, with layer number engineering, low cost and simple preparation, wet chemical processing, deterministic all-dry transfer process, surface passivation for dangle bond-free surface, band gap engineering, valleytronics, and so forth. Given these benefits, the preparation of 2DMs, specifically molybdenum disulfide (MoS_2), paves several advantages, and the use of near-infrared detection was the main topic of discussion in this chapter. MoS_2 is a well-known transition metal dichalcogenide (TMD) with a tunable bandgap (from ~1.8 eV in monolayer to ~1.2 eV in bulk). Heterostructures of MoS_2 enable band alignment optimization for extended NIR absorption. Further, the van der Waals heterostructures facilitate efficient charge transfer and reduce recombination losses. In addition, Type-II band alignment in MoS_2-based heterostructures promotes carrier separation, enhancing photocurrent generation. Due to these exceptional characters, the proposed chapter covers the fundamentals, operation principles, and fabrication of MoS_2 heterostructure-based photodetectors (heterojunction photodetectors), overcoming the generation of dark current, low quantum efficiency, and enhancing detectivity. This chapter offers a thorough guide to achieving high-performance near-infrared photodetectors based on MoS_2.

Keywords: MoS_2, NIR detection, van der Waals heterostructures, photogenerated carriers, carrier lifetime

1. Introduction

In the modern era of the electronics industry, photodetectors are essential parts. It is essentially an optoelectronic device that generates electricity from light. They are also called photosensors, which work primarily as the optical receivers to convert information from the light into electrical signals. They are crucial for capture, identification, and visualization of optical information [1]. The first photodetector emerged during World War II (1933) by Edgar Kutzscher, who found that lead sulfide (PbS) can be used as a photoconducting detector to detect hot emissions (infrared radiation) from bombers and ships. Our exploration of photo detectors began relatively recently, in the early 1990s [2]. In stark contrast, humanity's fascination with the nature of light stretches back to as early as 300 BC. This timeline shows our long-standing effort to understand light, laying the groundwork for the development of advanced photodetector technologies. The research on light started at 300 BC with Aristotle, followed by Huygens, who found the wave nature of light in the seventeenth century. Later in the eighteenth century, Maxwell predicted that those waves were electromagnetic waves with both electric and magnetic components. After many remarkable findings and theories in the nineteenth century, Einstein proposed that "light is composed of small packets of electromagnetic energy called photons." He even proved it with the help of the photoelectric effect, which is the basic principle behind the working of photodetectors [3]. Photodetectors work based on the principle of creating electron-hole pairs when exposed to light. When photons with energies equal to or greater than the material's bandgap hit the material, they excite electrons from the valence band to the conduction band, leaving behind positively charged holes. These conduction band electrons and valence band holes can move due to concentration gradients or electric fields, whether intrinsic or externally applied. These photogenerated electron-hole pairs may recombine and release light in the absence of an electric field. They can, however, be separated as free charge carriers by an electric field, and some of these free charges will gather at the electrode of the photodetector to form a photocurrent. This phenomenon of photocurrent generation is depicted in **Figure 1**. The strength of the incoming light has a direct correlation with the photocurrent's magnitude at a particular wavelength [4]. In 1800, Sir Frederick William Herschel paved the way to let us know about the wide spectrum of light. He was the first man to find that there is light beyond the red portion of the spectrum, and he named it "Calorific rays" which is now called "Infrared radiation." In 1801, followed by his finding, Johann Wilhelm Ritter found that there is also light beyond the violet portion, and he named it "Chemical rays" which is now called "Ultraviolet." Ultraviolet to infrared light is all the electromagnetic radiation with different wavelengths that constitutes the electromagnetic spectrum. The ability to detect light in different sub-bands has many uses in industries, the military, agriculture, and the biological sciences. Displays, video imaging, electronic eyes, optoelectronic storage, industrial safety, and other applications use visible (Vis) light detection, while ultraviolet (UV) light detection is used in military surveillance, space exploration, border security, environmental monitoring,

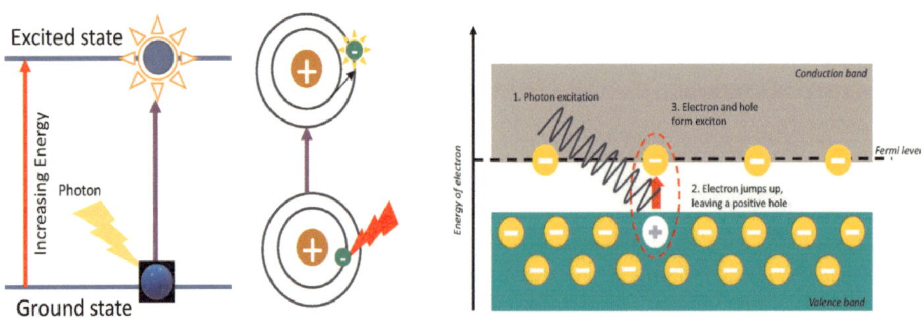

Figure 1.
The schematics of the operation principle of the photodetection (left), photon absorption and excitation (right), and electron-hole pair generation by the photoexcitation.

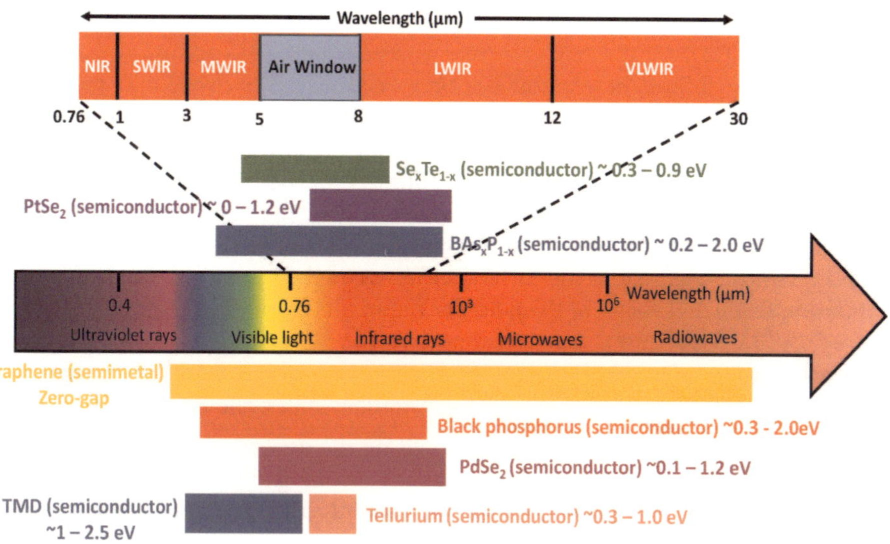

Figure 2.
Illustration of the range of energy of the electromagnetic spectrum with respective material classification.

and sterilization procedures, among other applications. Fire alarms, precision strikes, satellite remote sensing, optical communication, night vision technologies, remote control systems, analytical science, and navigational aids are just a few of the applications for infrared (IR) light detection [5]. **Figure 2** illustrates the broad electromagnetic spectrum, highlighting the wavelength ranges and the materials typically employed for detecting radiation in each spectral region. An ideal photodetector should efficiently convert light into electrical signals, detect weak optical signals, respond quickly, and utilize photons for charge carrier generation. These qualities are evaluated through figures of merit like responsivity, response time, noise equivalent power, detectivity, quantum efficiency, and gain, which are influenced by material properties such as bandgap, absorption coefficient, and charge carrier dynamics. Understanding these metrics helps assess material performance and optimization in photodetector applications. The formulas for each figure of merit and how they relate to material behavior and performance have been discussed in the following sections.

1.1 Responsivity

The ratio of the output photocurrent of the photodetector for a given incident optical power at a certain wavelength on the active region of the device.

$$R = \frac{I_{ph}}{P_{in} * A} = \frac{I_L - I_D}{P_{in} * A} \tag{1}$$

I_{ph} is the photocurrent, P_{in} is the incident optical power, I_L and I_D are the current under illumination and under dark, respectively, and A is the photosensitive area of the device. Responsivity is influenced by both the wavelength and intensity of the incident light. Its dependence on wavelength arises from the energy gap of the semiconductor and the wavelength-dependent nature of light absorption. This parameter characterizes the efficiency of photoelectric conversion in photodetectors and serves as a measure of the photocurrent generated under illumination at a specific power level [6].

1.2 Response time

It is typically defined in terms of rise time (τ_r) and fall or decay time (τ_f). The rise time corresponds to the time required for the photocurrent to increase from 10 to 90% of its peak value, while the fall time represents the duration needed for the photocurrent to decrease from 90–10% of its peak value [7]. In the frequency domain, the response time is associated with the bandwidth (Δf) of the photodetector, which determines the range of light modulation frequencies the device can effectively follow. Bandwidth is defined as the difference between the upper (f_2) and lower (f_1) modulation frequencies where the detector maintains a uniform response, expressed as $\Delta f = f_2 - f_1$ [8]. A commonly used parameter is the 3-dB bandwidth (f_{3dB}), the frequency at which the response drops to 50% of its peak value, related to the response time by $t_r = \frac{0.35}{f_{3dB}}$. For diode-based photodetectors, the response time is governed by the diffusion time of carriers exiting the depletion region, drift time within the depleted region, and junction capacitance. In contrast, photoconductors exhibit a response time dependent on carrier mobility and trap states, which can often be tuned through interface processing or modulation of trap states [9]. This parameter characterizes the ability of a photodetector to respond to rapid changes in light intensity and serves as a measure of the time required for the photocurrent to rise or fall between defined thresholds under modulated illumination.

1.3 Noise equivalent power (NEP)

It is described as the lowest optical power needed to obtain a signal-to-noise ratio (SNR) of 1 across a bandwidth of 1 Hz.

$$NEP = \frac{P_i}{\sqrt{\Delta f}} \tag{2}$$

Where P_i is the incident optical power that produces SNR = 1 and Δf is the frequency bandwidth [8]. Signal-to-noise ratio refers to the ratio of signal power and noise power in a photodetector. It is in units of decibels (dB) and is described as

$$SNR = \frac{signal\ power}{noise\ power} \tag{3}$$

A highly sensitive photodetector requires a high value of SNR [6]. In some cases, the NEP can be written as,

$$NEP = \frac{i_N}{R} \tag{4}$$

where i_N is the noise current spectra measured at 1 Hz bandwidth in darkness and R is the responsivity of the device.

$$i_N = \sqrt{\frac{1}{U}\int_0^U (I(t) - \langle I \rangle)^2\, dt} \tag{5}$$

Where $I(t)$ is the current at time t, and $\langle I \rangle$ is the average current over a period of U. Shot noise (i_{shot}), thermal noise ($i_{thermal}$), and flicker noise ($i_{flicker}$, $1/f$ noise) are the main contributors to the overall noise.

Shot noise is derived from random thermal agitation of charge carriers.

$$i_{shot} = \sqrt{2eI_D\Delta f} \tag{6}$$

where I_D is dark current and Δf is frequency bandwidth. Thermal noise, often referred to as Johnson-Nyquist noise, arises from the random motion of free carriers.

$$i_{thermal} = \sqrt{\frac{4k_BTB}{R_C}} \tag{7}$$

Where k_B is the Boltzmann constant, T is working temperature, and R_C is resistance. Flicker noise mainly occurs under low-frequency operation conditions.

$$i_{flicker} = K\frac{I_D^bB}{f^a} \tag{8}$$

Where K is a constant, f is the frequency of operation, and a and b are exponents characteristic of a specific device. NEP is determined by the intrinsic noise level, including contributions from thermal noise, shot noise, and flicker noise, which collectively define the minimum detectable signal. It is also influenced by the material properties, such as carrier mobility, recombination rates, and trap states [10]. A lower NEP indicates a higher sensitivity, as it quantifies its ability to detect weak optical signals in the total noise current spectrum [9].

1.4 Detectivity

It represents the sensitivity of the photodetector by characterizing its ability to distinguish and detect weak optical signals from noise.

$$D^* = \frac{\sqrt{A\,(\Delta f)}}{NEP} \tag{9}$$

Where A is the photosensitive area of the device and Δf is its bandwidth. When normalizing per square root of the area and the bandwidth of the photodetector, the detectivity is equal to the reciprocal of the Noise Equivalent Power. This parameter depends on the sensitivity, spectral response, and noise of the device. When the noise current is dominated by shot noise, detectivity can be calculated using

$$D^* = \frac{\sqrt{A}R}{\sqrt{2eI_d}} \tag{10}$$

Where I_d is the dark current and R is the responsivity. In the case of diode-like devices operating at zero bias, detectivity is expressed as

$$D^* = \frac{\sqrt{A}R}{2\sqrt{eI_{sat}}} \tag{11}$$

Where I_{sat} is the absolute value of the reverse saturation current. The unit of the detectivity is $cm.Hz^{1/2}/W$, which is also written as Jones. It is a useful parameter for comparing the detection performance of photodetectors with different materials and geometries. To reliably compare the metrics of performance, devices of the same sort and operating mode must be compared [9].

1.5 Quantum efficiency

Quantum efficiency measures the effectiveness of a photodetector in converting incident photons into charge carriers. The two classifications are external quantum efficiency (EQE) and internal quantum efficiency (IQE). EQE (η_{ext}) is defined as the ratio of the number of charge carriers contributing to the photocurrent (N_C) to the number of incident photons (N_I).

$$EQE\ (\eta_{ext}) = \frac{N_C}{N_I} = \frac{I_{ph}/e}{P/h\nu} = \frac{R.h.c}{e.\lambda} \tag{12}$$

Where I_{ph} is the photocurrent, e is the elementary charge, P is the incident optical power, h is Planck's constant, ν is the frequency, λ is the wavelength, c is the speed of light, and R is the responsivity. To achieve a high EQE, the active layer of the photodetector should exhibit strong optical absorption and longer carrier lifetimes relative to their transit time. This can be achieved by minimizing charge trapping and recombination, as well as addressing differences in electron and hole mobility. These factors enable carriers to contribute multiple times to the current before being collected [8]. IQE is the ratio of the number of electron-hole pairs contributing to the photocurrent to the number of photons absorbed by the detector. The absorbed photons are calculated by subtracting photon losses (from reflection and transmission) from the total incident photons.

$$IQE\ (\eta_i) = \frac{N_C}{N_A} = \frac{EQE}{\eta_A} \tag{13}$$

The number of absorbed photons (N_A) is calculated as $N_A = N_I * \eta_A$, where η_A is the light absorption efficiency. However, a photodetector cannot absorb all incident light due to reflection and transmission losses. The ratio of absorbed photons to incident photons is represented as light absorption efficiency (η_A) [7]. Additionally, some photogenerated electron-hole pairs recombine and are not collected as current, accounted for as collection efficiency (η_C).

In the absence of gain, EQE and IQE can be related as

$$\eta_{ext} = \eta_i * \eta_A * \eta_c \tag{14}$$

Here, η_i, η_A and η_c are less than 1, and η_{ext} is always smaller than η_i. For high-quality photodiodes, IQE can approach 1, meaning all absorbed photons generate electron-hole pairs. EQE and IQE provide complementary insights, where EQE includes all processes from photon incidence to current generation, while IQE focuses on the efficiency of photon absorption and charge generation within the active layer. In practical applications, achieving high EQE involves optimizing the active layer for efficient light absorption, reducing carrier recombination, and improving charge collection efficiency. For ultrathin photodetectors, it is also essential to account for optical interference effects that influence photon absorption. A thorough understanding of EQE and IQE is crucial for designing photodetectors with enhanced sensitivity and performance.

1.6 Gain

It evaluates the capability of the photodetector to generate multiple charge carriers from a single incident photon. It quantifies the extent of photo-exciton recycling through a multiplication mechanism and denotes the number of charge carriers detected per incident photon.

$$G = \frac{I_{ph}/e}{\eta P_{in}/h\nu} = R\frac{hc}{\eta e\lambda} \tag{15}$$

Where c is the velocity of light, λ is the excitation wavelength, and η is quantum efficiency. If there are no losses from recombination or other inefficiencies, the photodetector's gain is directly proportional to both the IQE and EQE [11].

The gain can also be expressed as

$$G = \frac{\tau_e}{t_{tre}} + \frac{\tau_p}{t_{trp}} \tag{16}$$

τ_e and τ_p are the lifetimes of photogenerated electrons and holes, respectively, while t_{tre} and t_{trp} are the transit times between electrodes for electrons and holes, respectively. In the case of a photodiode, the lifetime of the photogenerated carriers is equal to their transit time, and the gain is 1. The relation between gain and EQE is given by

$$\eta_{ext} = \eta_A G \tag{17}$$

The EQE for a photodiode is typically less than 100% when no carrier multiplication is occurring. In the case of an ideal photodiode with an EQE of 100%, the responsivity is expressed as,

$$R = \frac{\lambda}{1.24} \tag{18}$$

1.6.1 Gain in photoconductors

In photoconductors with a perfect semiconductor, the gain arises due to the difference in mobility between electrons and holes. Electrons, having higher mobility, move faster under an external field. When an electron-hole pair is generated in the device's center, the photogenerated electron is collected at the drain electrode before the hole reaches the source electrode. To maintain charge neutrality, an electron is injected from the source electrode. This cycle continues until the hole reaches the source electrode or recombines with an electron, allowing multiple electrons to be collected for each photon, resulting in gain. In imperfect semiconductors, gain can also result from charge traps. The photogenerated holes get trapped by these defect traps. In order to maintain the charge neutrality, electrons have to be injected from the source electrode, which will be collected at the drain. When a photon causes more than one electron to be collected, then gain will occur.

1.6.2 Gain in photodiodes

Photodiodes operate by separating and collecting photogenerated carriers using the built-in electric field. The injection of electrons into the p-type semiconductor is not possible because of its minority nature despite being more mobile than holes. For holes, the p-type semiconductor and electrode make an Ohmic contact; for electrons, it creates a blocking contact, which will also lead to the restriction of electrons. For these reasons, even in the reverse bias condition, no gain is obtained.

Without permitting carrier multiplication, the reverse bias voltage merely increases the depletion region and the electric field inside it, increasing light absorption and carrier drift velocity. Gain cannot be induced by charge traps in the depletion zone because carrier injection is prevented by the blocking contact. Instead, these traps may cause recombination of photogenerated carriers, reducing the photo response. To achieve a gain in photodiodes, alternative carrier multiplication mechanisms are needed. One such mechanism is avalanche breakdown, which involves impact ionization. This process leads to carrier multiplication and gain, which is widely employed in avalanche photodiodes. However, avalanche breakdown requires a high reverse bias. Additionally, photogenerated hot carriers with sufficient energy can also cause carrier multiplication, resulting in external quantum efficiencies (EQE) greater than 100%. The impact ionization can be obtained only when the energy of the photon surpasses the bandgap energy [9]. Silicon (CMOS-based) photodetectors dominate the photodetector market due to their outstanding photoelectric characteristics, affordability, and seamless integration with electronic systems. However, Si cannot absorb the majority of the infrared spectrum due to its bandgap of about 1.1 eV [12]. In addition, due to its indirect bandgap nature, Si has a relatively weak absorption. Such inherent properties of Si make it challenging to develop active photodetectors. Recent years have seen a bottleneck in the fabrication of silicon-based devices, which would result in the end of Moore's Law [13]. To address these limitations, researchers have focused on integrating silicon with other semiconductor materials such as germanium; Group III-V semiconductors: gallium arsenide (GaAs), indium phosphide (InP), gallium nitride (GaN) etc.; Group II-VI semiconductors: cadmium telluride (CdTe), cadmium sulfide (CdS), etc.; and two-dimensional materials. Over the past decade, there has been considerable effort to design and fabricate photodetectors that combine silicon and various semiconductor materials. This strategy has earned significant attention due to its potential to improve the performance of silicon-based photodetectors. However, these materials had drawbacks such as limited flexibility, irreversible dangling bonds, a short bandgap, and low optical transparency. In their quest for superior materials, scientists have discovered the promising potential of two-dimensional (2D) materials in the realm of photodetectors. These materials, with their unique properties, have opened new chapters for innovation, offering enhanced performance and capabilities in light detection technologies. Since the isolation of graphene (Gr), by Novoselov and Geim in 2004, 2D materials (2DMs) have evolved rapidly during the last 15 years. With hundreds of members, 2DMs have so far expanded into a large family encompassing graphene, group VA semiconductors, group IVB transition metal dichalcogenides (TMDCs), group VIB TMDCs, group IVA metal chalcogenides (MCs), black phosphorus (b-P), hexagonal boron nitride (h-BN), etc. Two-dimensional (2D) materials are inherently layered, making them outstanding candidates for photodetectors. Their high charge carrier mobility, combined with the extraordinary tunability of the band gap through variations in thickness, external strain, and electric fields, sets them apart [6]. Additionally, these materials can be easily stacked to form van der Waals heterostructures, effectively reducing lattice mismatch. These unique characteristics position 2D materials as superior choices for cutting-edge photodetector technology. Graphene, the first two-dimensional (2D) material to gain attention for photodetection applications, owes its prominence to its exceptional electrical properties. Notably, its remarkable planar mobility, which reaches up to 200,000 $cm^2/(V{\cdot}s)$, facilitates the creation of photodetectors with bandwidths as high as 40 GHz. However, one major limitation of graphene as an active layer in photodetectors is its lack of an energy band gap, which results in significant noise contributions from dark currents [14]. Consequently, the scientific community has increasingly focused on 2D materials with finite band gaps. Among the array of two-dimensional materials, transition metal dichalcogenides (TMDCs) have gained significant attention. Their unique properties, such as the transition from an indirect to a direct band gap when their dimensions are reduced from the bulk material to the monolayer limit, and this dimensional reduction enhances their light-matter interaction and tunable electronic characteristics, make them stand out in the field of photodetectors. These materials promise enhanced performance, with intriguing electro-optical properties despite their relatively modest mobility of 200 $cm^2/(V{\cdot}s)$. TMDCs exhibit absorbance values that are an order of magnitude higher than those of silicon (Si) and gallium arsenide (GaAs), enabling efficient light absorption with a very thin layer of the photoactive material. These properties position TMDCs as promising candidates for advanced photodetector applications, offering significant advantages over traditional materials. Among the TMDCs, a group VI transition metal dichalcogenide, molybdenum disulfide (MoS_2) attracted much interest in the last decade for photodetection because of its efficient absorption and emission of light. MoS_2 structure consists of 3 atomic layers, where a layer of transition metal atoms (Mo, W) is positioned between two layers of chalcogen atoms (S, Se, Te) with strong in-plane covalent and weak out-of-plane van der Waals interactions, resulting in easy exfoliation into monolayers with atomically thin thickness [15]. The molybdenite is relatively abundant in nature; this widespread availability of molybdenite makes MoS_2 an economically viable option for large-scale production, ensuring a steady supply of raw material. MoS_2 has a tunable bandgap that changes from 1.3 eV for multilayer (ML) MoS_2 (indirect bandgap) to 1.9 eV for single layer (SL). This change in bandgap value is arising from d-orbital-related interactions [16, 17]. MoS_2 exhibits a high absorption coefficient of up to 2.8×10^6 cm^{-1}, which is at least one order of magnitude higher than that of conventional semiconductors like Si and GaAs. This is a reflection of the strong light-matter interaction that distinguishes MoS_2 (19). Two separate low-energy peaks in the MoS_2 absorption spectra at 1.88 eV (A excitons) and 1.06 eV (B excitons) were obtained because of the direct and indirect transitions at the K point. The PL peaks appear approximately at 620 and 688 nm, corresponding to the indirect-gap and direct-gap luminescence. MoS_2 monolayers exhibit exceptional photoluminescence quantum yield (PLQY), reaching over 95%. Additionally, they have a remarkable minority carrier lifetime, with the longest-observed duration being 10.8 ± 0.6 nanoseconds by means of surface passivation [18, 19]. Van Hove singularities in the electronic density of states of TMDC guarantee enhanced light-matter interactions, leading to enhanced photon absorption and electron-hole creation (which are collected in transparent graphene electrodes). This allows development of extremely efficient flexible photovoltaic devices with photoresponsivity above 0.1 ampere per watt (corresponding to an external quantum efficiency of above 30%) [20]. In ultrathin 2D MoS_2, reduced electronic Coulomb screening due to charge confinement significantly increases exciton binding energy. This enhancement alters the wavelengths of light absorbed or emitted, resulting in an exciton binding energy increase of hundreds of meV. These characteristics position MoS_2 as a leading candidate for photodetection applications [21].

2. Fabrication techniques of MoS_2 heterostructure photodetectors

The concept of assembling and fabricating 2D materials, including MoS_2, into heterostructures was first proposed by Wang et al. in their work on high-quality graphene electronics [22], leading to the discovery of numerous novel properties in these engineered systems. These heterostructures can be broadly classified into two categories: vertically stacked heterostructures, where layers are assembled via van der Waals interactions, and epitaxially grown planar heterostructures, where materials are seamlessly integrated through direct growth, ensuring strong in-plane bonding. Among the various methods of fabrication of these heterostructures, the Chemical Vapor Deposition (CVD) and deterministic transfer methods are the powerful, efficient way to attain the better devices with remarkable properties. These methods were briefly discussed here. A high-precision method for creating MoS_2 heterostructures, the deterministic transfer method guarantees regulated monolayer stacking free from contamination or misalignment. The schematic experimental setup and the deterministic transfer process have been illustrated in the **Figure 3(a)** and **(b)** respectively. The development of high-performance van der Waals (vdW) heterostructures, where exact interlayer coupling directly affects electronic and optical properties, requires this technique. First, bulk crystals of monolayer or few-layer MoS_2 are usually exfoliated by applying liquid exfoliation, Nitto tape, or scotch tape on SiO_2/Si, sapphire, or other substrates. A viscoelastic polymer, like polydimethylsiloxane (PDMS) or poly (methyl methacrylate) (PMMA), is applied to the monolayer as a support layer during the transfer process. With the aid of micromanipulators and an optical microscope, the MoS_2 flake is precisely positioned over the target substrate or another two-dimensional material (e.g., WS_2, graphene). A temperature-controlled stamp-assisted technique ($\approx$ 40–100°C) can be used to achieve the clean adhesion while reducing interfacial contamination. Ultimately, the pristine interface is restored by annealing (approximately 300–400°C in Ar/H_2 atmosphere) or acetone removal of the polymer layer [23].

Chemical Vapor Deposition (CVD) is a widely used bottom-up approach for the fabrication of vertical 2D heterostructures, enabling the growth of planar multi-junction heterostructures. Several strategies have been explored to achieve scalable and controllable fabrication of both vertical and planar heterostructures with atomically sharp heterojunctions and clean interfaces. The two-step CVD method, commonly used for this purpose, involves growing layered crystals as a substrate for the second layer. The rate of gas flow and synthesis time play crucial roles in the

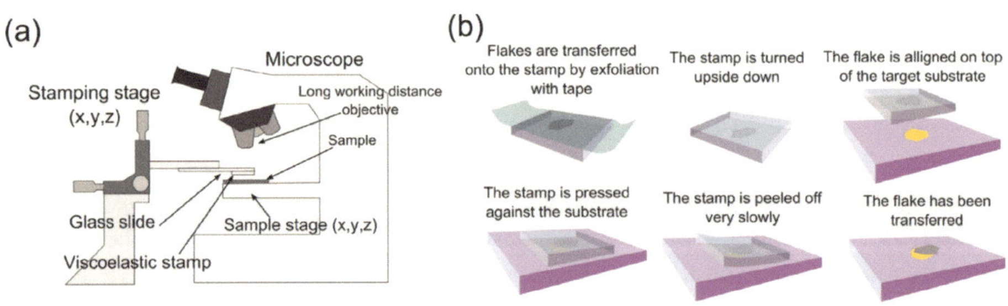

Figure 3.
Deterministic setup and procedure for transfers. (a) schematic representation of the experimental configuration used for the all-dry transfer procedure. The viscoelastic stamp preparation and the deterministic transfer of an atomically thin flake onto a user-defined location (such as another atomically thin flake) are depicted in the diagram (b) reproduced with permission from [23]

formation of vertically stacked and lateral heterostructures. For instance, the growth of a 2D GaSe/MoSe2 heterostructure was achieved using this method, though large lattice misfits between the layers can result in incommensurate superstructures [24]. Additionally, a stacked TMD/hBN heterostructure was synthesized using a Ni-Ga alloy and Mo foil as substrates, where the alloy promoted the formation of the hBN honeycomb lattice while Mo served as a source of Mo [25]. The multi-step CVD method offers greater flexibility and control by switching the direction and modulating the components of the gas flow. This approach allows for sharper heterojunction boundaries and enables sequential growth of multi-junction heterostructures. For example, a $WS_2 - WS_2(1 - x)$ Se2x monolayer planar heterostructure with tunable band alignment was successfully grown by modulating the gas flow direction and temperature. Similarly, a one-pot synthesis strategy was developed by placing two precursor powders (MoX_2 and WX_2) in the same boat, with different gas flows promoting the selective growth of MoX_2 and WX_2, respectively [26]. This multi-step approach, compared to one-step or two-step CVD methods, offers more precise control, enabling the creation of spatially distinct optoelectronic devices by separating electrons and holes into different materials. Dong Hee Shin et al. fabricated MoS_2/SQD:SiO_2 photodiodes by direct chemical vapor deposition (CVD) of MoS_2 onto SiO_2-embedded silicon quantum dot (SQD) multilayers. The SQD:SiO_2 multilayers were first prepared by reactive ion beam sputtering, followed by high-temperature annealing at 1100°C in a nitrogen atmosphere to form well-embedded quantum dots. High-resolution transmission electron microscopy (HRTEM) and photoluminescence (PL) confirmed the presence of SQDs. For MoS_2 deposition, SQD: SiO_2 substrates were placed downstream of MoO_3 powder, and hydrogen sulfide (H_2S) was introduced to synthesize single- and multilayer MoS_2 films at 600°C. The MoO_3 density was controlled to regulate the film thickness, which was analyzed using Raman spectroscopy. Atomic bonding states of MoS_2 were characterized using X-ray photoelectron spectroscopy (XPS), while its optical absorption was evaluated using a UV-Vis-NIR spectrophotometer. To enhance photodetector (PD) performance, a bathocuproine (BCP) layer was spin-coated onto the Si surface, serving as an electron-selective contact to minimize leakage current. Finally, Au and InGa electrodes were deposited using photolithography and a lift-off process. This fabrication method ensures high-quality MoS_2 growth with minimized defects at the MoS_2/SQD:SiO_2 interface, leading to enhanced carrier transport. The resulting photodetector achieves exceptional performance, including a high external quantum efficiency (EQE) of 92%, a detectivity (D*) of 6.1×10^{13} Jones, and an ultrafast response time (60 ns rise/756 ns fall), making it one of the fastest MoS_2-based PDs reported. The integration of SQDs further boosts light absorption and carrier extraction, making this fabrication approach highly effective for high-speed and broadband photodetection **Figure 4** [27].

The MoS_2/GaN/Si heterostructure photodetector (PD) was fabricated by Deependra Kumar Singh et al. to achieve self-powered, broadband, and spectrally distinctive photodetection. A 250 nm GaN thin film was first grown on Si(111) substrates using plasma-assisted molecular beam epitaxy (PAMBE). The GaN layer plays a crucial role in forming a heterojunction with MoS_2, enabling efficient carrier transport and built-in potential formation. Following GaN deposition, MoS_2 thin films were grown using pulsed laser deposition (PLD). An MoS_2 pellet (2 cm diameter) served as the target, positioned 4.5 cm from the substrate. The deposition was performed at a base pressure of 5.0×10^{-6} mbar, using a pulsed laser source with a repetition rate of 3 Hz, 1000 laser pulses, and a laser energy of 200 mJ, resulting in a laser

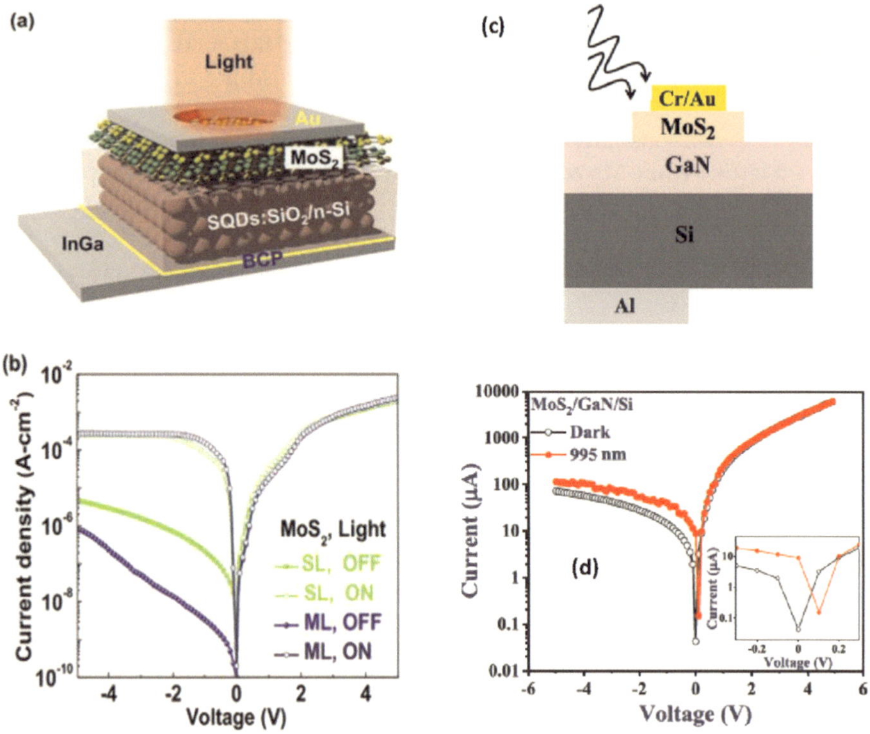

Figure 4.
(a). A schematic of a typical MoS_2/SQDs:SiO_2 MLs/n-Si PD with Au and InGa as top and bottom electrodes, respectively. (b)Dark/photo J-V curves at λ = 500 nm in the range of bias from 5 to 5 V. (1) (c) Schematic of the device indicating the direction of light illumination. (d) Dark and illuminated I-V characteristics of the MoS_2/GaN/Si-based device (inset shows the I-V characteristics near zero bias) reproduced with permission from [27].

fluence of 1 J/cm^2. During deposition, the chamber pressure was maintained at 1.5×10^{-5} mbar. The sample underwent in situ annealing at 700°C for 20 minutes, followed by slow cooling at a ramp rate of 5°C/min to enhance film crystallinity. This fabrication approach enables the realization of a high-performance photodetector with a broadband response (300–1100 nm) and an ultrahigh responsivity of 23.81 A/W at 995 nm under low-intensity illumination (0.075 mW/cm^2) in a self-powered mode. The device exhibits a unique feature of photocurrent polarity inversion in the NIR region due to a competitive mechanism between the photothermal electric (PTE) effect in MoS_2 and the built-in potential at the MoS_2/GaN/Si heterojunction. The demonstrated wavelength-dependent polarity switching opens new possibilities for low-power optoelectronic applications, making this PD one of the most advanced MoS_2-based devices for future filter-less and energy-efficient photodetection technologies [28].

Minkyun Son et al. fabricated MoS_2/HfS_2 heterojunction photodetector. Firstly, MoS_2 was synthesized via a two-step CVD process in a 2-inch tubular furnace. A SiO_2/p-Si substrate was pre-deposited with MoO_3 and NaCl using thermal evaporation, then placed in a crucible at the center of the furnace, while 4 g of sulfur was positioned 20 cm upstream. The system was heated at 30°C/min under 200 sccm Ar flow at 760 Torr, held at 150°C for 15 min, and then heated at 25°C/min to 500°C. At this point, sulfur vaporization began, and when the chamber temperature reached 750°C, the sulfur heating zone reached 200°C, ensuring uniform MoS_2 growth.

The temperature was maintained for 20 min before natural cooling. The grown MoS_2 exhibited large flake sizes (~200 μm) with 86% coverage, providing a high-quality platform for heterostructure fabrication. The HfS_2/MoS_2 vertical heterojunction was synthesized in a 1-inch tubular furnace with MoS_2/SiO_2 placed at the center and a mixture of $HfCl_4$ (0.02 g) and NaCl (0.01 g) positioned further away in a crucible. Upstream, 4 g of sulfur was kept 15 cm from MoS_2, with H_2 (15 sccm) and Ar (20 sccm) as carrier gases. The system was heated at 30°C/min to 150°C (held for 15 min), followed by a faster ramp at 40°C/min to 500°C, after which the sulfur heating zone was activated. When the main heater reached 950°C, sulfur reached 150°C, maintaining these conditions for 30 min before natural cooling. This controlled synthesis enabled the selective growth of HfS_2 on MoS_2 due to strong interfacial interactions, forming a vertical heterojunction with a large interface area, enhancing interlayer exciton generation. The conduction band edge of HfS_2 (5.2 eV) resulted in a narrow interlayer bandgap, facilitating IR detection. The built-in potential at the interface suppressed dark current and created gradient band bending, ensuring efficient carrier drift. The resulting HfS_2/MoS_2 photodetector demonstrated exceptional performance, achieving a photoresponsivity of ≈600 A/W and detectivity (D^*) of $\approx 7 \times 10^{13}$ Jones at 1550 nm and ≈1500 A/W with $D^* \approx 2 \times 10^{14}$ Jones at 980 nm. The device exhibited rapid response times (rise/decay: 60/71 μs), surpassing conventional MoS_2-based detectors. This interlayer exciton-driven IR photodetector, fabricated via a scalable two-step CVD process, marks a significant advancement in high-performance 2D optoelectronic devices **Figure 5** [30].

Subhrajit Mukherjee et al. synthesized MoS_2 quantum dots (QDs) via solvent-assisted sonochemical exfoliation, modified to ensure uniform film deposition on Si for vertical Si/MoS_2 p-n heterojunction fabrication. Since high-boiling DMF caused QD aggregation, prolonged sonication was followed by 2 days of vigorous stirring for size uniformity. Gradual centrifugation yielded QD precipitates of different sizes, which were dried under Ar at ~60°C and redispersed in ethanol for spin-coating onto H-passivated p-type Si (0.77 Ω-cm). Device fabrication included Au (~70 nm) as the top electrode and Al (~80 nm) on the Si backside via thermal evaporation (base pressure $\sim 1 \times 10^{-6}$ Torr). For LED fabrication, a ~ 20 nm Al-doped ZnO (AZO) film was deposited via pulsed laser deposition (PLD) at ~300°C (base pressure $\sim 5 \times 10^{-6}$ Torr, energy density ~ 2 J/cm^2, repetition rate ~ 5 Hz), exhibiting a resistivity of $\sim 10^{-3}$ Ω·cm and ~ 92% transmittance. MoS_2 QDs exhibited size-dependent emission (2–26 nm) with nearly stoichiometric 2H-phase and n-type doping due to Cl, confirmed via XPS. Time-resolved PL showed superior carrier lifetimes (1.5–2.5 ns) compared to MoS_2 flakes (~10 ps). The 0D/3D heterostructures demonstrated rectifying characteristics (ideality factor ~ 9.0) and bias-dependent electroluminescence (450–800 nm), indicating potential for LED applications. Spectral responsivity increased with decreasing QD size, reaching 0.85 A/W and a peak detectivity of $\sim 8 \times 10^{11}$ Jones at −2 V for ~2 nm QDs, outperforming commercial Si homojunctions and graphene/Si devices. The study highlights the potential of colloidal n-MoS_2 QDs for Si-compatible, large-area multifunctional optoelectronic devices using 0D/3D heterojunctions [29].

3. MoS_2 heterostructure for near IR detection

Our recent work has employed metal passivation by Zn and Fe on MoS_2 to achieve high-performance in the NIR regime and low dark current. As seen in **Figure 6d**, the

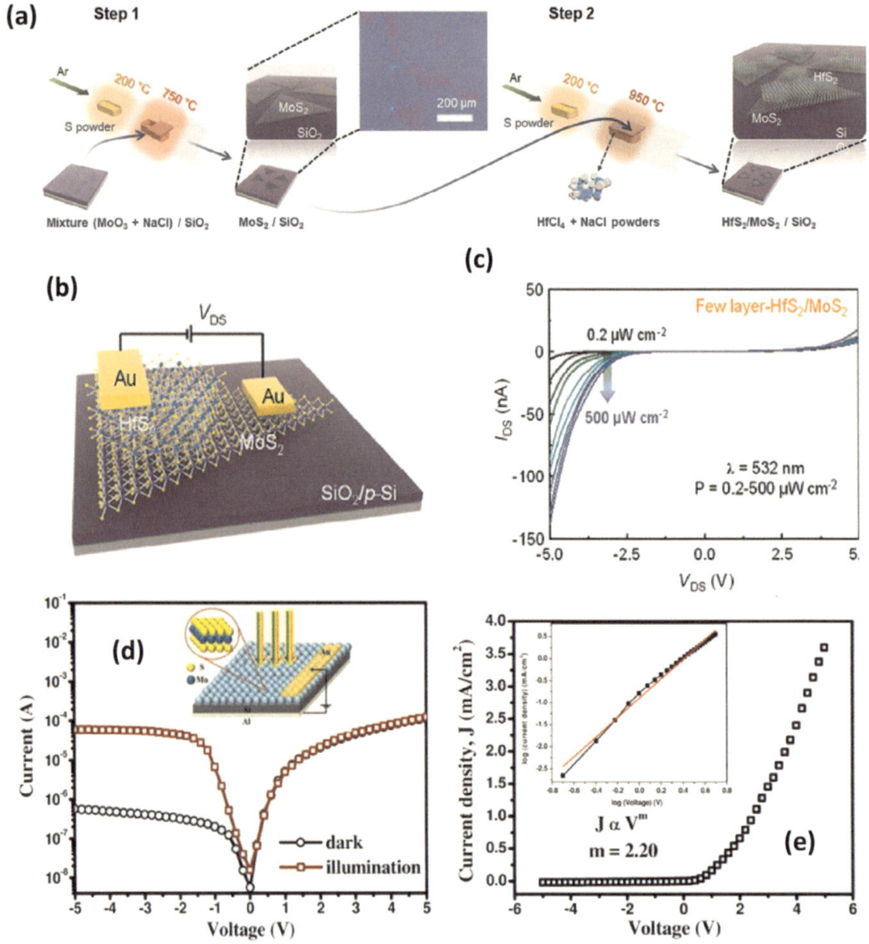

Figure 5.

(a). Schematic of overall two-step HfS_2/MoS_2 synthesis procedures. The inset image is of synthesized MoS_2 and the scale bar denotes 200 µm. (b) Schematic of HfS_2/MoS_2 photodetector. (c) IDS versus VDS of the HfS_2/MoS_2 photodetector in the dark and at a different P (λ = 532 nm). (3) (d) I-V characteristics of fabricated heterojunction, using 2 nm MoS_2 QDs on Si, under dark and illumination conditions. (e) J-V characteristics of fabricated p-n heterojunctions under dark conditions, at room temperature. The inset shows the logarithmic J-V curve for the heterojunction fitted with a power law J .Vm corresponding to the trap-charge-limited current transport. Reproduced with permission from Ref. [29].

synthesized pure MoS_2 (MoS), Zn-MoS, and Fe-MoS are tested in both light and dark conditions. The thermally persuaded charge carriers, also known as leakage current, are typically responsible for the dark current (I_{dark}), which is measured at $1.5x10^{-6}$ A, $3.98x10^{-6}$ A, and $2.81x10^{-6}$ A for MoS, Zn-MoS, and Fe-MoS devices, respectively. Ag (metal electrode)-MoS_2 (semiconductor) photodetectors with leakage current (I_{leak}) exhibit Schottky behavior in dark conditions, as demonstrated by the obtained I-V curve. The carrier injection, which regulates the flow of carriers into semiconductors, is directly impacted by the metal-semiconductor interface. Surface passivation of the (Zn and Fe)-MoS_2 surface, which reduces the surface states, could counteract the fermi level pinning (FLP) effect that would deteriorate the carrier injection through the metal-semiconductor (M-S) interface. Band bending (uplift/downshift) is facilitated by the gap states formed by the metal contact between the semiconductor energy bands or by crystal defects or distortion that induce the surface states. Additionally, in order to prevent leakage, the FLP effect is suppressed by altering the surface of MoS_2 with Fe and Zn, which passivates the dangling bonds on the surface and allows for the establishment of a downward shift in fermi energy. Therefore, as illustrated in **Figure 6d**, the enhanced photoresponse of the devices is indicated by the increased current (I_{illum}) of 1.89x10^{-6} A, 9.77x10^{-6} A, and 6.25x10^{-6} A for MoS, Zn-MoS, and Fe-MoS under the light illumination. I_{ph} = I_{dark} - I_{illum} is used to calculate the photocurrent (I_{ph}) at a bias voltage of 3 V. For MoS, Zn-MoS, and Fe-MoS, the enhanced I_{ph} values are 0.3x10^{-6} A, 5.79x10^{-6} A, and 3.44x10^{-6} A, respectively. The trapping mechanism can be clearly understood from the temporal photoresponse measurement. **Figure 6a** shows non-exponential behavior in the light ON–OFF condition in the I_{Illum} -t curve, indicating. The trapping of carriers induced photocurrent. For 10 seconds, the MoS first cycle maintains a dark steady-state current of 1.59x10^{-6} A. Light is illuminated in such a situation, and an increase in current (1.89x10^{-6} A) is observed with a similar steady state for 10 s. The behavior returns with a similar trend as seen in the previous cycle when the dark and light conditions are switched. Comparably, the dark current for (Zn and Fe)-MoS is 3.98x10^{-6} and 2.81x10^{-6} A, respectively, and it abruptly rises to 9.77x10^{-6} and 6.25x10^{-6} A when illuminated. **Figure 6** illustrates the response's improved stability across various cycles (a, b, and c). For MoS, Zn-MoS, and Fe-MoS, the computed rise time (τ_r) and decay time (τ_d) are 0.8, 1.3, and 1 s, and 1, 1.2, and 1.5 s, respectively [31]. This is what causes the photocurrent's abrupt increase and rapid decay. In another work by Vabbina et al. designed and fabricated a shortwave infrared (SWIR) photodiode by integrating a few layers of p-type molybdenum disulfide (MoS_2) onto a graphene layer, forming a van der Waals (vdW) heterojunction. This device exhibited an extensive spectral response covering the SWIR region, as shown in **Figure 7b**. The photodetector achieved a peak responsivity of 1.26 A/W at a wavelength of 1.44 µm, with a corresponding specific detectivity of 4 × 10^{10} Jones. The broadband spectral response of this photodetector was attributed to the coexistence of two distinct operational mechanisms: the optical bandgap excitation mode and the internal photoemission mode. In optical bandgap excitation mode, when the energy of incident photons is greater than the intrinsic bandgap of MoS_2, electron-hole pairs are generated, resulting in a strong

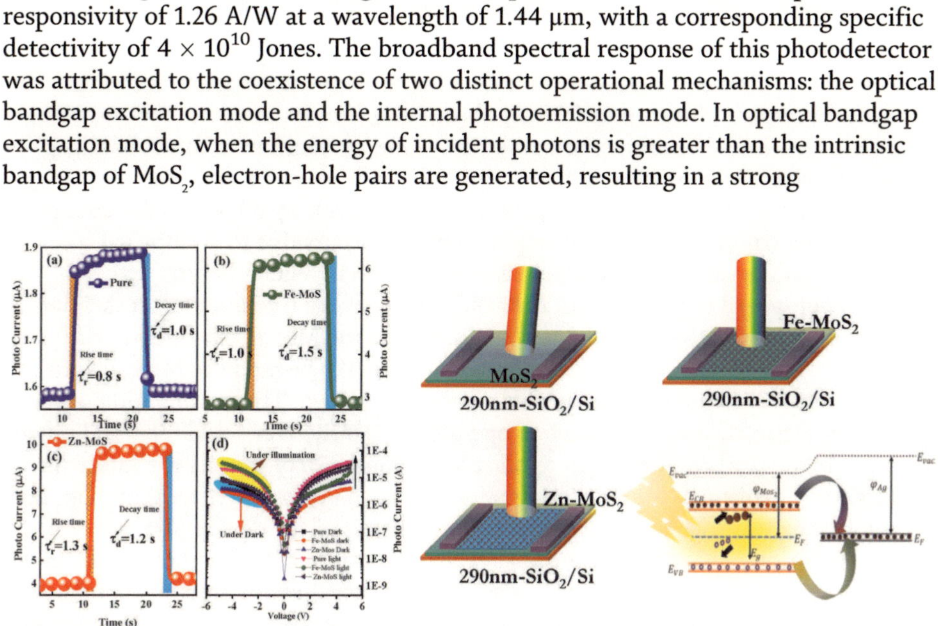

Figure 6.
Temporal photocurrent measurement for pure MoS_2 *(a) and metal-passivated* MoS_2 *photodetectors with (b) Fe, (c) Zn, and (d) photocurrent of pure, Zn and Fe passivated MoS2 photodetectors [31].*

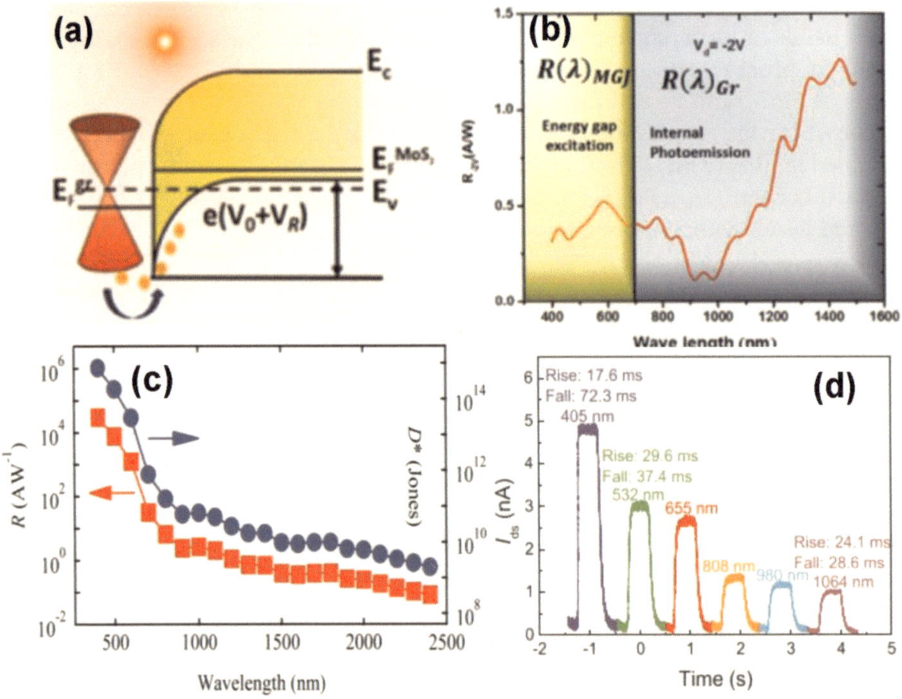

Figure 7.
(a) Band alignment of graphene/MoS_2 heterostructure panel; (b) responsivity as a function of the wavelength of the incident light at a bias of -2 V for IR photodiode; (c) responsivity and the specific detectivity variation for different wavelengths of the incident illumination; and (d) transient frame of source-drain current under illumination with various wavelengths [32].

photocurrent. This mechanism is responsible for the device's response at shorter wavelengths, where MoS_2 can efficiently absorb and convert incident photons into charge carriers. In the internal photoemission mode, in addition to direct bandgap excitation, the photodetector can also operate through an internal photoemission process. In this mode, electron-hole pairs are excited even by photons with energy lower than the bandgap of MoS_2. This phenomenon occurs due to an internal transition mechanism between the graphene and MoS_2 layers. When photons with sub-bandgap energy strike the junction, they induce electronic transitions that facilitate photoemission across the potential barrier formed at the MoS_2/graphene interface. This process enables the detection of longer wavelengths beyond the MoS_2 absorption edge, significantly extending the device's spectral response. The combination of these two modes allows the photodetector to operate efficiently over a wide spectral range, making it highly suitable for SWIR applications. **Figure 7a** illustrates the internal transition mechanism and the resulting photoemission across the heterojunction, further explaining the broadband response characteristics of the device [32].

Long et al. introduced an advanced shortwave infrared (SWIR) photodiode based on a MoS_2/graphene/WSe_2 van der Waals (vdW) heterojunction, in which a monolayer of graphene was sandwiched between a bottom p-type MoS_2 layer and a top n-type WSe_2 layer, all deposited onto a SiO_2/Si substrate. This unique layered architecture enabled the photodiode to leverage the distinct optical and electronic properties of each material, leading to enhanced performance in terms of broadband absorption and carrier transport. One of the primary advantages of this device configuration was the inclusion of a graphene interlayer, which contributed significantly to broadband light absorption [33]. Unlike conventional semiconductors with a fixed bandgap, graphene, being a zero-bandgap material, possesses a continuous density of electronic states that allows it to absorb a wide spectrum of incident light, from visible to infrared wavelengths. This characteristic was instrumental in extending the photodetection range of the device far beyond that of MoS_2 or WSe_2 alone. The incorporation of MoS_2 (p-type) and WSe_2 (n-type) formed a well-defined p–n junction, which played a crucial role in the efficient separation and transport of photogenerated charge carriers. Upon illumination, electron-hole pairs were generated at different regions of the heterostructure. The built-in electric field at the MoS_2/WSe_2 junction facilitated the rapid separation of these charge carriers, reducing recombination losses and improving the overall quantum efficiency of the photodetector.

The combination of broadband absorption in graphene and the efficient charge separation mechanism in the p–n heterojunction allowed the device to achieve a remarkably wide spectral response. The photodetector exhibited sensitivity across a broad wavelength range, spanning from 0.4 μm (visible) to 2.4 μm (SWIR), as illustrated in **Figure 7c**. Furthermore, the synergistic effects of the vdW heterojunction led to an improvement in specific detectivity, enabling the device to operate with high sensitivity even in low-light conditions. The MoS_2/graphene/WSe_2 vdW heterojunction photodiode demonstrated a significant advancement in optoelectronic device engineering by integrating the unique properties of 2D materials to achieve high-performance broadband photodetection. Its ability to detect wavelengths from the visible to the SWIR region, coupled with high carrier mobility and efficient charge separation, makes it highly suitable for applications in optical sensing, imaging, and telecommunications. More notably, the MoS_2/graphene/WSe_2 vdW heterojunction photodiode exhibited an impressive responsivity of approximately 2.64×10^3 A/W at a forward bias of 1 V when illuminated with 1.064 μm wavelength light. This high responsivity was accompanied by a remarkable specific detectivity of 1.1×10^{13} Jones and an external quantum efficiency (EQE) reaching nearly 10^4. An EQE exceeding unity indicates the presence of a photoconductive gain in the device, which can be attributed to several key mechanisms.

First, the short transit distance in the Vertical Junction: The vertical charge transport between MoS_2, graphene, and WSe_2 allowed photogenerated carriers to traverse the junction efficiently, reducing recombination losses. Then, the interlayer inelastic tunneling effect is the presence of graphene as an intermediate layer that facilitated charge transfer across the heterojunction through inelastic tunneling, further enhancing carrier collection efficiency, and finally, the rapid carrier transport, which is the high carrier mobility within the vdW heterostructure, enabled efficient charge separation and extraction, contributing to the significant photoconductive gain. Additionally, the photodetector demonstrated a fast response time across various wavelengths. For instance, at 1.064 μm, the device exhibited a rise time of 24.1 ms, as shown in **Figure 7d**. This rapid response, combined with its high responsivity and detectivity, underscores the potential of the MoS_2/graphene/WSe_2 heterojunction photodiode for high-performance optoelectronic applications, including infrared imaging and optical communication [36]. Long et al. demonstrated the fabrication of a mid-wave infrared (MWIR) photodiode by utilizing the narrow optical bandgap and the ease of heterojunction formation in two-dimensional black arsenic-phosphorus (b-AsP), specifically in the form of As_xP_{1-x}. The resulting b-AsP/MoS_2 stacked van der Waals (vdW) heterojunction photodiode exhibited outstanding MWIR photodetection capabilities. Under broadband illumination in the 3–5 μm range, the device achieved an exceptional specific detectivity exceeding 4.9×10^9 Jones at zero bias and room temperature. This performance significantly surpassed that of certain commercially available infrared photodetectors, including PbSe-based and thermistor-based IR photodetectors, as illustrated in **Figure 8a**. In conclusion, the b-AsP/MoS_2 vdW heterojunction photodiode demonstrated superior MWIR photodetection performance, achieving a high specific detectivity of 4.9×10^9 Jones at zero bias and room temperature. Its broadband response in the 3–5 μm range outperformed conventional IR photodetectors, highlighting the potential of 2D materials for next-generation infrared sensing applications. The exceptional specific detectivity of the b-AsP/MoS_2 vdW heterojunction photodiode was primarily attributed to its suppressed dark current, resulting from the potential barrier at the p-type b-As_xP_{1-x} and n-type MoS_2 heterojunction, as illustrated in **Figure 8 (b)** and **(c)**. This barrier effectively reduced noise, leading to a two-order magnitude improvement in detectivity compared to a bare b-As_xP_{1-x} photoconductor. In addition to its high detectivity, the photodiode exhibited an ultrafast response time of approximately 0.5 ms at 4 μm, as shown in **Figure 8b** [37]. Several black phosphorus (bP)-based vdW heterojunction photodiodes, such as the MoS_2/bP heterojunction for 1.55 μm detection, have also shown excellent performance in the shortwave infrared (SWIR) region in addition to MWIR detection [26], the WSe_2/bP/MoS_2 sandwich-like heterojunction for 1.55 μm [38], and

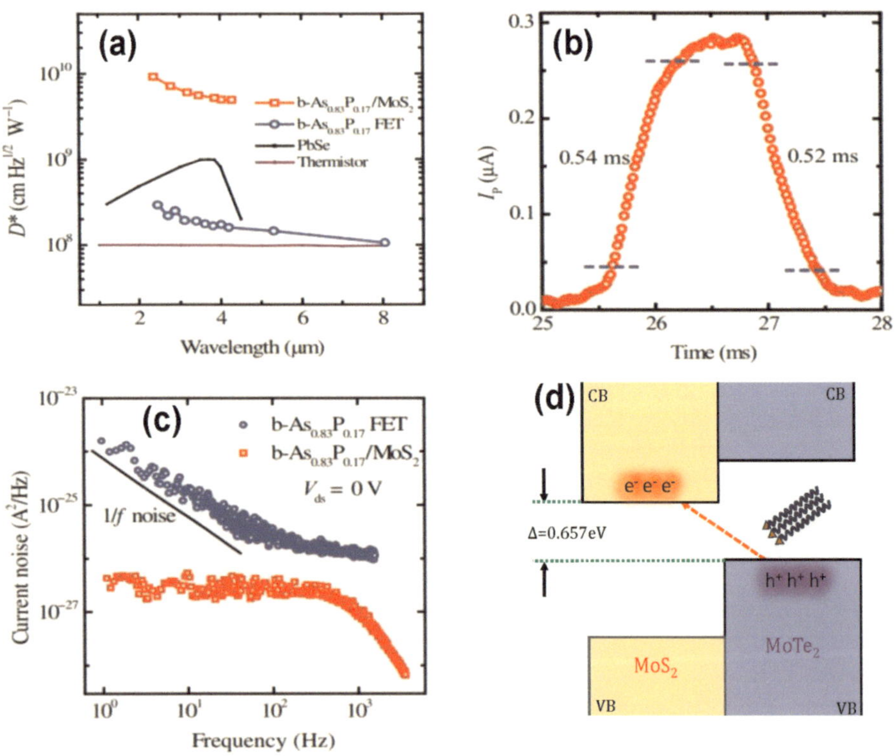

Figure 8.
(a) Specific detectivity as a function of the wavelength of incident light for b-AsP/MoS_2 photodiode; (b) time transient response behavior of 4 μm illumination; (c) current noise and frequency for the IR photodetector based on the heterojunction [34]; and (d) band diagram of the $MoTe_2$/MoS_2 heterostructure photodiode [34].

the $PdSe_2$/bP heterojunction for 1.31 μm detection [35]. These results highlight the versatility of bP-based vdW heterojunctions for broadband infrared photodetection. 2D layered materials-based heterojunctions and CNT-based heterojunctions were fabricated, and both of their properties were compared, from which it is proven that the problems of organic NIR detectors and CNT-based NIR detectors, such as environmental instability and low sensitivity, are said to be overcome with the help of layered materials-based heterojunctions. The graphene/MoS_2 heterojunction is fabricated using a two-step process in which a single layer of graphene is grown on the Cu foil by the chemical vapor deposition method, and then it is transferred to the SiO_2/Si substrate by patterning it as an 800 nm nanoribbon using electron beam lithography, and then MoS_2 flakes are exfoliated and transferred onto the graphene channel by means of PMMA-mediated transfer technique. Finally, source and drain electrodes were also constructed using electron beam lithography. Further, the $WeSe_2$/MoS_2 heterojunction is fabricated by means of mechanical exfoliation and micromanipulation methods. WSe_2 was transferred to the SiO_2/Si substrate after getting exfoliated from the crystal. MoS_2 flake was placed on WSe_2 by means of a micromanipulation

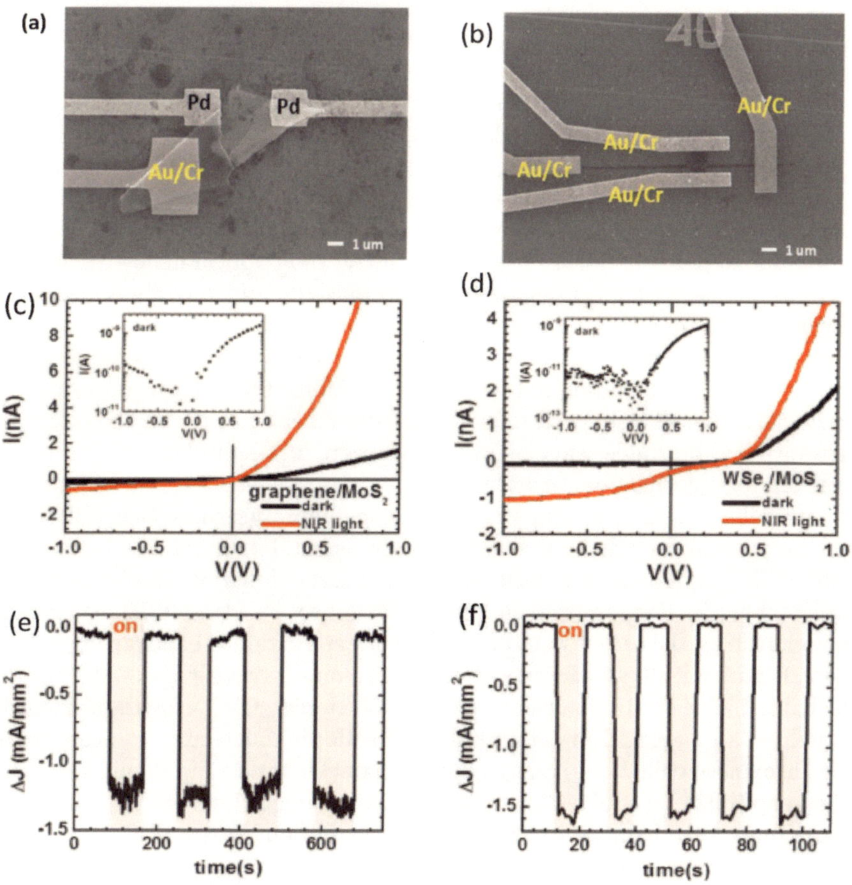

Figure 9.
Field emission scanning electron microscope images of the graphene/MoS_2 (a), WSe2/MoS_2 (b), graphene/MoS_2 (c), and WSe2/MoS_2 (d) I-V curves measured across the junction under NIR light and in the dark are displayed in the insets on semi-logarithmic scales. When a laser diode (λ = 808 nm) with a power of 5.3 mW mm^{-2} was turned on and off repeatedly for the graphene/MoS_2 (e), WSe2/MoS_2, the temporal photoresponse was measured at V = -0.5 V (f) [35].

method with the aid of polydimethylsiloxane film and a 3-axis manipulator. Finally, Au/Cr and Pd electrodes were constructed on MoS_2 and WSe_2 flakes, respectively, using electron beam lithography and lift-off techniques (**Figure 9a** and **b**). Both the heterojunctions show the diode rectifying behavior (forward bias) in the dark, and the photocurrent was increased when the heterojunctions were illuminated with NIR light (785 cw laser), which shows its response in the NIR region. (**Figure 9c** and **d**) The inset figure represents the on/off current ratio at V = ±1 V, which was about 10 and 200, respectively. In particular, WSe_2/MoS_2 heterojunction exhibited photovoltaic effect. The photocurrent density is measured at V = - 0.5 V for both the heterojunctions by using the laser diode of wavelength 808 nm which is turned on and off repeatedly (**Figure 9e** and **f**). The photocurrent rise and decay times were shorter than 1 s, which makes it better than CNT/ZnO heterojunction with the rise and fall times of 90 and 70 s, respectively [35].

4. Summary and outlook

2D material-based infrared (IR) photodetectors have emerged as promising candidates for next-generation optoelectronic devices due to their exceptional properties, including tunable bandgaps, high carrier mobility, and strong light-matter interactions. These photodetectors can be classified into three main categories based on device configurations: photoconductors, photodiodes, and nanophotonic-integrated photodetectors. Photoconductors, particularly graphene-based ones, offer high-speed performance, while photodiodes exhibit improved photoresponsivity and noise reduction due to p–n junctions. Additionally, integrating optical nanostructures enhances light absorption, leading to higher responsivity and detectivity. Despite significant advancements, several challenges include limited spectral coverage: most 2D material-based IR photodetectors operate in the SWIR and MWIR ranges, with only a few materials extending to LWIR/FIR, crucial for thermal imaging and biosensing. The Incomplete performance characterization in various photodetector evaluations focus on single-wavelength illumination, lacking full spectral response data needed for optimization. In addition to this, the real-world applicability concerns rely on laser illumination, making practical deployment uncertain. Using blackbody illumination would provide a more realistic assessment. Finally, the scalability issues in current fabrication methods, primarily mechanical exfoliation, hinder large-area and uniform production, limiting commercial viability. To advance 2D material-based IR photodetectors, the area that requires further exploration is bandgap engineering through 2D material alloys—adjusting material composition (e.g., InxGa1 − xP alloys) could enable LWIR to FIR detection. Further, developing CVD growth methods for high-quality, large-area 2D materials tailored for IR photodetectors and comprehensive performance evaluation, i.e., systematic characterization across multiple wavelengths to optimize spectral response. Remarkably, integration with optical nanostructures unambiguously enhances light absorption via waveguides, cavities, plasmonic structures, and large-scale detector arrays, which address integration challenges for high-resolution imaging applications requiring CMOS-compatible processing. By overcoming these challenges, 2D material-based IR photodetectors could revolutionize imaging, sensing, and optoelectronic technologies, paving the way for high-performance, scalable, and broadband photodetection systems.

Author details

Vinoth Elangovan[1], Naskath Sithara H[2,3], Harish Santhana Krishnan[1], Navaneethan Mani[1] and Jayaram Archana[2,3]*

1 Nanotechnology Research Center, Faculty of Engineering and Technology, SRM Institute of Science and Technology, Chennai, Tamil Nadu, India

2 Centre of Excellence in Materials and Advanced Technologies (CeMAT), Faculty of Engineering and Technology, SRM Institute of Science and Technology, India

3 Functional Materials and Energy Devices Laboratory, Department of Physics and Nanotechnology, SRM Institute of Science and Technology, Chennai, Tamil Nadu, India

*Address all correspondence to: jayaram.archana@gmail.com

References

[1] Chee KWA. Introductory chapter: Photodetectors. In: Advances in Photodetectors. London, UK: IntechOpen; 2019

[2] Rogalski A. History of infrared detectors. Opto-Electronics Review. 2012;**20**:279-308

[3] Yajnik UA. The conception of photons - Part I. Resonance. 2015;**20**: 1085-1110

[4] Donati S. Photodetectors: Devices, Circuits and Applications. 2nd ed. IEEE Press Wiley; 2021

[5] Romoli M et al. The ultraviolet and visible-light coronagraph of the HERSCHEL experiment. AIP Conf. Proc. 2003;**679**:846-849

[6] Wang F et al. 2D library beyond graphene and transition metal dichalcogenides: A focus on photodetection. Chemical Society Reviews. 2018;**47**:6296-6341

[7] Long M et al. Progress, challenges, and opportunities for 2D material-based photodetectors. Advanced Functional Materials. 2018;**28**:1803807

[8] Xie C et al. Photodetectors based on two-dimensional layered materials beyond graphene. Advanced Functional Materials. 2016;**27**:1603886

[9] Wei Y et al. Recent advances in photodetectors based on two-dimensional material/Si heterojunctions. Applied Sciences. 2023;**13**(19):11037

[10] Yao J et al. 2D material broadband photodetectors. Nanoscale. 2020;**12**: 454-476

[11] Wang J et al. Design strategies for two-dimensional material photodetectors to enhance device performance. InfoMat. 2019;**1**:251-290

[12] Wadhwa R et al. A strategic review of recent progress, prospects and challenges of MoS_2-based photodetectors. Journal of Physics D: Applied Physics. 2022;**55**:063002

[13] Cheng Y et al. Research progress on improving the performance of MoS_2 photodetector. Journal of Optics. 2022; **24**:105001

[14] Bernardi M et al. Optical and electronic properties of two-dimensional layered materials. Nano. 2017;**6**(2): 479-493

[15] Li H et al. From bulk to monolayer MoS_2: Evolution of Raman scattering. Advanced Functional Materials. 2012;**22**: 1385-1390

[16] Tian H et al. Optoelectronic devices based on two-dimensional transition metal dichalcogenides. Nano Research. 2016;**9**:1543-1560

[17] Ahmad S et al. A comparative study of electronic properties of bulk MoS_2 and its monolayer using DFT technique: Application of mechanical strain on MoS_2 monolayer. Graphene. 2014;**3**(4):73-81

[18] Kwak JY et al. Absorption coefficient estimation of thin MoS_2 film using attenuation of silicon substrate Raman signal. Results in Physics. 2019;**13**: 102202

[19] Amani M et al. Near-unity photoluminescence quantum yield in MoS_2. Science. 2015;**350**(6264): 1065-1068

[20] Britnell L et al. Strong light-matter interactions in heterostructures of

atomically thin films. Science. 2013; **340**(6138):1311-1314

[21] Chaves A et al. Bandgap engineering of two-dimensional semiconductor materials. npj 2D Materials and Applications. 2020;**4**:29

[22] Dean CR, Young AF, Meric I, Lee C, Wang L, et al. Boron nitride substrates for high-quality graphene electronics. Nature Nanotechnology. 2010;**5**(10): 722-726

[23] Castellanos-Gomez A, Buscema M, Molenaar R, Singh V, Janssen L, van der Zant HSJ, et al. Deterministic transfer of two-dimensional materials by all-dry viscoelastic stamping. 2D Materials. 2014;**1**:011002

[24] Gong YJ, Lin JH, Wang XL, Shi G, Lei SD, et al. Vertical and in-plane heterostructures from WS2/MoS_2 monolayers. Nature Materials. 2014; **13**(12):1135-1142

[25] Fu L, Sun YY, Wu N, Mendes RG, Chen LF, et al. Direct growth of MoS_2/h-BN heterostructures via a sulfide-resistant alloy. ACS Nano. 2016;**10**(2): 2063-2070

[26] Zheng BY, Ma C, Li D, Lan JY, Zhang Z, et al. Band alignment engineering in two-dimensional lateral heterostructures. Journal of the American Chemical Society. 2018; **140**(36):11193-11197

[27] Shin DH et al. High-speed heterojunction photodiodes made of single- or multiple-layer MoS_2 directly-grown on Si quantum dots. Journal of Alloys and Compounds. 2020;**820**: 153074

[28] KumarSingh D et al. Differentiation of ultraviolet/visible photons from near infrared photons by MoS_2/GaN/Si-based photodetector. Applied Physics Letters. 2021;**119**:121102

[29] Mukherjee S et al. Novel colloidal MoS_2 quantum dot heterojunctions on silicon platforms for multifunctional optoelectronic devices. Scientific Reports. 2016;**6**:29016

[30] Son M et al. High-performance infrared photodetectors driven by interlayer exciton in a Van Der Waals epitaxy grown HfS_2/MoS_2 vertical heterojunction. Advanced Optical Materials. 2023;**34**:2308906

[31] Abinaya R et al. Modulating Fermi energy in few-layer MoS_2 via metal passivation with enhanced detectivity for near IR photodetector. Journal of Materials Chemistry C. 2024;**12**: 5247-5254

[32] Vabbina P et al. Highly sensitive wide bandwidth photodetector based on internal photoemission in CVD grown p-type MoS_2/graphene Schottky junction. ACS Applied Materials & Interfaces. 2015;7:15206-15212

[33] Long M et al. Broadband photovoltaic detectors based on an atomically thin heterostructure. Nano Letters. 2016;**16**(4):2254-2259

[34] Long M et al. Room temperature high-detectivity mid-infrared photodetectors based on black arsenic phosphorus. Science Advances. 2017; **3**(6):e1700589

[35] Park MJ et al. Near-infrared photodetectors utilizing MoS_2-based heterojunctions. Journal of Applied Physics. 2015;**118**:044504-044510

[36] Ye L et al. Near-infrared photodetector based on MoS_2/black phosphorus heterojunction. ACS Photonics. 2016;**3**(4):692-699

[37] Li H et al. High-performance broadband floating-base bipolar phototransistor based on WSe_2/BP/MoS_2 heterostructure. ACS Photonics. 2017; **4**(4):823-829

[38] Afzal AM et al. High-performance p-BP/n-$PdSe_2$ near-infrared photodiodes with a fast and gate-tunable photoresponse. ACS Applied Materials & Interfaces. 2020;**12**(17):19625-19634

Chapter 4

Understanding Photodiodes through Simulation: A Case Study of GaAs PIN Photodiodes

Badra Bouabdallah, Saliha Kheris and Zakia Nabi

Abstract

Photodiodes remain a key focus in various fields of electronics due to their remarkable properties. These properties are primarily determined by the materials used in their fabrication. Gallium arsenide (GaAs) is a promising material for photodiode manufacturing and serves as an excellent model for simulating transport phenomena in these structures. Notably, the presence of trapping centers significantly impacts their performance, as detailed in this chapter. This chapter presents both physical and mathematical models that exploit the properties of GaAs and the phenomena observed in a PIN diode under dark conditions. The study initially focuses on the behavior of the diode in the absence of illumination. However, the chapter also incorporates the physical processes associated with the photovoltaic mechanism into the mathematical approach, providing a simplified model to govern and predict phenomena under illumination. The basic operation of the PIN photodiode is explained, followed by an analysis of the physical phenomena occurring in dark conditions, including the role of trapping centers in recombination and space charge storage mechanisms. Furthermore, the chapter distinguishes between conduction simulations in a standard PIN diode and a PIN photodiode. Finally, a simple analytical model derived from simulation results is introduced to explore the electronic behavior. Detailed physical insights are provided to better understand the observed phenomena and validate the model.

Keywords: GaAs, PIN photodiode, modelization, Gummel-Newton method, simulation

1. Introduction

The research topic of this chapter is centered around the modeling and simulation of optoelectronic and electronic components, specifically focusing on the advantages offered by Gallium Arsenide (GaAs) and its impact on improving component performance. GaAs is known for its superior material properties, making it highly suitable for high-speed and high-sensitivity applications in optoelectronics. The primary focus is on understanding the transport phenomena within devices based on PN junctions, which is a core concept for the development of efficient semiconductor devices. These phenomena are often simulated using software tools that integrate simplified physical models to predict the behavior of charge carriers in the device [1].

As the size of electronic components continues to decrease significantly, the need for accurate prediction models becomes even more critical. These models must evolve alongside advancements in technology, which involves both refining existing models and developing new physical models. Furthermore, rapid implementation of these models in commercial software is crucial for practical applications. Often, functional programming tools such as MATLAB, Mathematica, and Mathcad are used because they are more user-friendly and less complex than traditional programming languages like C++ and Pascal, which require deeper technical knowledge.

Recent decades have seen considerable progress in understanding diffusion phenomena in semiconductors, including the behavior of free charge carriers and doping in diodes. Multiple models have been proposed to describe these phenomena, with a particular focus on GaAs due to its unique properties. One critical aspect when modeling GaAs is the presence of trap centers that can affect the transport of charge carriers. These traps must be carefully accounted for in simulations to accurately predict the performance of devices based on GaAs. A comprehensive physical model that incorporates these complexities can lead to more accurate simulations and, ultimately, devices with better performance [2, 3].

In addition to the advancements in understanding the transport phenomena, the field of wide-bandgap semiconductors (with an energy gap greater than 2.5 eV) has emerged as a new frontier in semiconductor physics and technology. These materials have vast potential in optoelectronics, particularly in applications like blue and ultraviolet light emission, as well as in high-temperature and high-power microelectronics. The development of wide-bandgap materials, such as III-V compounds, II-VI compounds, and IV-IV compounds, has revolutionized the industry by providing new possibilities for creating efficient optoelectronic components. The ability to form ternary and quaternary alloys from these materials further expands their applications, allowing for precise tuning of their electronic and optical properties [4, 5].

The study of electronic and optical properties, as well as the computational models of key physical quantities, is crucial for advancing semiconductor technology. This chapter explores the generation and recombination processes, transport mechanisms in the presence of trap centers, and the fundamental equations that arise from these physical models. By understanding these effects, it becomes possible to develop more effective and efficient components [3].

The focus of this research is on understanding the interaction mechanisms between charge carriers and external forces such as electric and magnetic fields. This knowledge is critical for optimizing the performance of semiconductor devices. A detailed study of GaAs and similar materials will allow us to better understand how external conditions, such as temperature and applied fields, influence the behavior of charge carriers and, ultimately, the efficiency of devices. To achieve this, it is necessary to model the movement of individual electrons, considering the interactions between electrons and lattice imperfections. This requires advanced computational techniques that simulate the probabilistic nature of these interactions [4].

The goal is to create a simple yet accurate computational model that can be applied to a wide range of materials and semiconductor devices. By doing so, we can develop tools that can be widely used for designing and optimizing future electronic and optoelectronic components [1, 2].

2. The system GaAs

GaAs is widely used in components for ultrafast electronics, optoelectronics, optical communication, and microwave applications [2].

The direct bandgap of GaAs gives rise to interesting electronic properties [6]. The effective mass of electrons in the central valley (Γ) is very low [7], resulting in high electron mobility [8]. The bandgap energy is relatively high at 1.424 eV at 300 K [7, 9], compared to 1.12 eV for silicon [7, 10] and 0.66 eV for germanium [7, 11]. At low electric fields, electron mobility in GaAs is higher than in silicon, even at high doping levels [12]. Combined with the effects of velocity saturation in GaAs [13], this leads to reduced transit times [14].

GaAs is a semi-insulating material. Experimental research has demonstrated the existence of deep centers due to doping [4, 5, 15, 16], as shown in the **Figure 1**, where several peaks are clearly visible and can be identified.

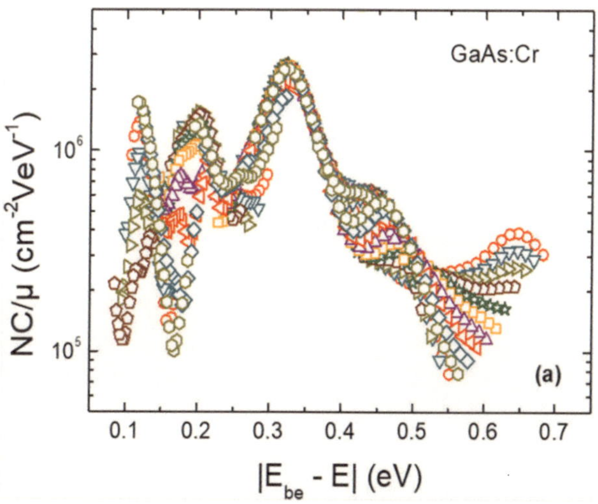

Figure 1.
Modulated Photocurrent Spectroscopy (MPC) of the density of states in the bandgap of a GaAs:Cr crystal [17].

2.1 Optical properties of GaAs

When exposed to light, photons with energy Eph greater than the bandgap energy (Eph > Eg) can excite electrons from the valence band to the conduction band. This process may occur either without phonon involvement (where the electron's wave vector k remains conserved)—case of GaAs—or with phonon participation (where the phonon momentum is added to that of the electron). These transitions result in the absorption of light via band-to-band transitions, leading to the generation of excess electron-hole pairs.

Figure 2 shows the absorption spectra of chromium-doped GaAs, calculated using the Wien2k code [18].

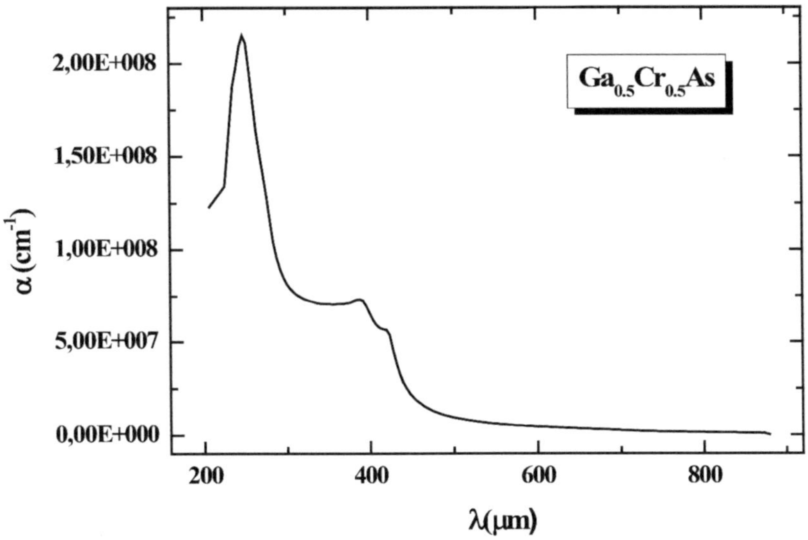

Figure 2.
Calculated absorption spectra of GaAs:Cr [18].

2.2 Optical activity in crystals

An optically active crystal creates two circular polarizations of incident light in opposite directions, propagating at different speeds [19]. In GaAs, birefringence can be induced by an electric field applied perpendicular to the direction of light propagation. When the electric field is applied along the <100> crystallographic direction, the refractive index change Δn associated with this direction is proportional to the intensity of the applied field [20].

$$\Delta n = \frac{n^3 r}{2} \xi\langle 100\rangle \quad (1)$$

where:

- r is the electro-optical coefficient,
- n is the refractive index,
- ξ is the applied electric field intensity.

For GaAs at 300 K, the values are:

- $r = 1.6 \times 10^{-10} \mathrm{cm \cdot V^{-1}}$
- n = 3.34
- $n^3 r = 59 \times 10^{-10} \mathrm{cm \cdot V^{-1}}$

The following sections present some fundamental physical concepts necessary for the approach we have adopted.

3. Conduction in the presence of two deep centers

The semiconductor GaAs exhibits a deep donor center E_T (EL2) level and an acceptor center N_A in the band gap [21–23], where the deep donor center compensates the acceptor center [24].

Many experimental results suggest that the recombination mechanism is controlled by a third center called the recombinant center E_R [25]. located near the middle of the bandgap, as shown in the band diagram in **Figure 3**.

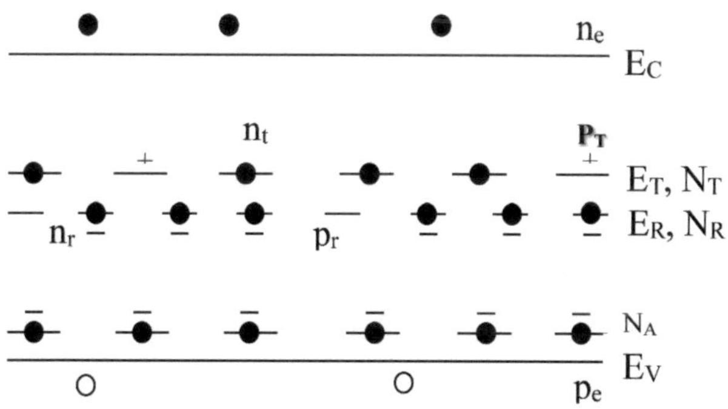

Figure 3.
Descriptive band diagram of GaAs semi-insulator.

This figure should depict the energy band diagram of a GaAs semi-insulator, illustrating the following key features:

- *Conduction Band (CB):* The highest energy band where electrons can freely move.
- *Valence Band (VB):* The lowest energy band where holes can freely move.
- *Bandgap (E_g):* The energy difference between the conduction band and the valence band.
- *Fermi Level (E_F):* The energy level at which the probability of finding an electron is 50%. In a semi-insulator, the Fermi level is typically located within the bandgap.
- *Donor Level (EL2):* The energy level of the deep donor impurity (EL2) within the bandgap.
- *Acceptor Level (N_A):* The energy level of the acceptor impurities within the bandgap.
- *Recombination Center (E_R):* The energy level of the recombination center located near the mid-gap.

The whole equations are related and corresponds to semiconctor material's physics [26–33].

The concentration N_A actually represents an effective concentration:

$$N_A = N_A + N_{D\ (\text{shallow})} \tag{2}$$

The system of equations that expresses the transport mechanism is written as:

$$\begin{aligned}
\frac{\partial n}{\partial t} &= \frac{1}{e}\,\text{div}\,\overrightarrow{J_n} - U_{nr} - U_{nt}\\
\frac{\partial p}{\partial t} &= -\frac{1}{e}\,\text{div}\,\overrightarrow{J_p} - U_{pr} - U_{pt}\\
\frac{\partial n_r}{\partial t} &= U_{nr} - U_{pr}\\
\frac{\partial n_t}{\partial t} &= U_{nt} - U_{pt}
\end{aligned} \tag{3}$$

Where U is the net recombination rate, r is for recombinant, and t is for trap.

Assuming an exchange mechanism of the Shockley-Read type between the deep centers N_R (or N_T) and the bands:

$$\begin{aligned}
U_{nr} &= C_n^R\left[n.p_r - n_{lr}.n_r\right]\\
U_{nt} &= C_n^T\left[n.p_t - n_{lt}.n_t\right]\\
U_{pr} &= C_p^R\left[p.n_r - p_{lr}.p_r\right]\\
U_{pt} &= C_p^T\left[p.n_t - p_{lt}.p_r\right]
\end{aligned} \tag{4}$$

In these equations, n_r and p_r are the electron and hole densities at the recombinant center, which is assumed to be of acceptor type (neutral if the center is empty and negatively charged if it is occupied by an electron). Similarly, n_t and p_t are the electron and hole densities at the deep center EL2, which is assumed to be of donor type.

We have then:

$$\begin{aligned}
n_r + p_r &= N_R\\
n_t + p_t &= N_T
\end{aligned} \tag{5}$$

The shallow acceptor center is assumed to be completely ionized in the vicinity of room temperature, and it is assumed that the concentrations of N_R and N_T are sufficiently low to neglect the inter-center exchange. At thermal equilibrium ($U \cong 0$), we have the following relationships for each center:

$$\begin{aligned}
n_1 &= \frac{e_n}{C_n}\\
p_1 &= \frac{e_p}{c_p}
\end{aligned} \tag{6}$$

then:

$$\begin{aligned}
n_{1r} &= \frac{n_e \cdot p_{re}}{n_{re}}\, p_{1r} = \frac{p_e \cdot n_{re}}{p_{re}}\\
n_{1t} &= \frac{n_e \cdot p_{te}}{n_{te}}\, p_{1t} = \frac{p_e \cdot n_{te}}{p_{te}}
\end{aligned} \tag{7}$$

We have also:

$$\tau_{nre} = \frac{1}{C_n^R.N_R}\tau_{pre} = \frac{1}{C_p^R.N_R} \qquad \tau_{nte} = \frac{1}{C_n^T.N_T}\tau_{pte} = \frac{1}{C_p^T.N_T} \tag{8}$$

3.1 At steady state

In steady state, $U_{nr} = U_{pr}$ and $U_{nt} = U_{pt}$, which, combined with Eq. (5), leads to:

$$C_n^R[n.p_r - n_{lr}.n_r] = C_p^R[p.n_r - p_{lr}.p_r] \qquad C_n^T[n.p_t - n_{lt}.n_t] = C_p^T[p.n_t - p_{lt}.p_t] \tag{9}$$

Where:

$$p_r = N_R\frac{\tau_{pre}.n_{lr} + \tau_{nre}.p}{\tau_{pre}.(n + n_{lr}) + \tau_{nre}.(p + p_{lr})} \qquad n_r = N_R\frac{\tau_{pre}.n + \tau_{nre}.p_{lr}}{\tau_{pre}.(n + n_{lr}) + \tau_{nre}.(p + p_{lr})} \tag{10}$$

and:

$$p_t = N_T\frac{\tau_{nte}.p + \tau_{pre}.n_{lt}}{\tau_{pte}.(n + n_{lt}) + \tau_{nte}.(p + p_{lt})} \qquad n_t = N_T\frac{\tau_{nte}.p_{lt} + \tau_{pte}.n}{\tau_{pte}.(n + n_{lt}) + \tau_{nte}.(p + p_{lt})} \tag{11}$$

4. Test structure

The studied structure is a P + -ν-N+ photodiode, as shown in **Figure 4**. The base material is N-type GaAs, chosen for its high density of traps (N_T) [24, 34]. The analysis of transport mechanisms is focused on the region near the P + -ν contact. The length of the n-zone is sufficiently long to establish ohmic conduction in the central part of the volume and to prevent the space charge region from extending into either contact.

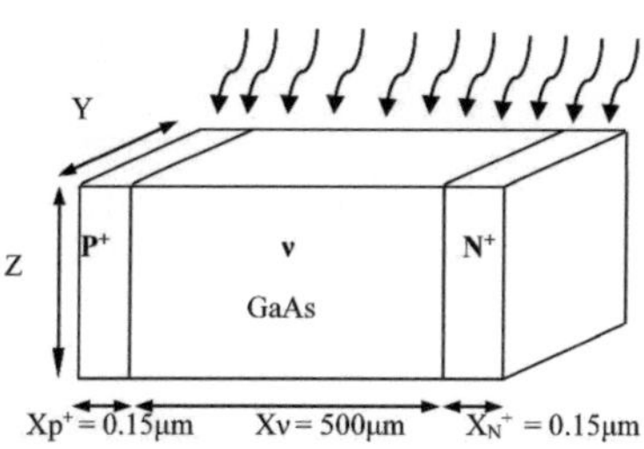

Figure 4.
P^+-ν-N^+ test structure.

A center involved in the Shockley-Read recombination mechanism [25] is characterized by four independent parameters:

- n_t *and* p_t *(in* cm^{-3}*)*: These represent the electron and hole densities at the trap level in thermal equilibrium, which depend on the energy position of the trap level within the bandgap.

- τ_{nte} *and* τ_{pte} *(in* s^{-1}*)*: These are the electron and hole capture lifetimes at the trap level, which are related to the capture cross-sections of the trap for electrons and holes, respectively.

- N_T *(in* cm^{-3}*)*: This denotes the density of recombination centers in the material.

Table 1 presents the relevant physical parameters of the test structure [10, 26, 34].

Parameters	Symbols	Value	Unit
Effective density of states in conduction band	N_c	4.43×10^{17}	cm^{-3}
Effective density of states in valence band	N_v	8.84×10^{18}	cm^{-3}
Relative permittivity	E_r	12.5	—
Intrinsic carrier concentration	n_i	2×10^6	cm^{-3}
Bandgap energy	E_g	1.432	eV
Electron mobility	μ_n	4000	$cm^2/V.s$
Hole mobility	μ_p	280	$cm^2/V.s$
EL2 trap:			
Trap density	N_T	2×10^{16}	cm^{-3}
Electron capture time constant	T_{nt}	1.5×10^{-6}	s
Hole capture time constant	T_{pt}	2.7×10^{-9}	s
Recombination Center (Er):			
Electron capture time constant	τ_{nr}	10^{-10}	s
Hole capture time constant	τ_{pr}	10^{-10}	s
Electron density at thermal equilibrium	n_{1r}	4.45×10^5	cm^{-3}
Hole density at thermal equilibrium	p_{1r}	8.989×10^6	cm^{-3}
General parameters:			
Lifetime	τ_0	1.56×10^{-10}	s
Debye length	L_{Da}	0.34	mm
Electron density at thermal equilibrium (EL2)	n_{1t}	2.1×10^6	cm^{-3}
Hole density at thermal equilibrium (EL2)	p_{1t}	1.905×10^6	cm^{-3}

Table 1.
The relevant physical parameters of the test structure.

5. Photodetection characteristics

Any photon does not automatically undergo photodetection in optical, optoelectronic, or telecommunication systems [35, 36].

First, the photon must have an energy E_{ph} equal to or greater than the band gap energy Eg to allow the transition of the electron from the valence band to the conduction band. This implies a cutoff wavelength λc above which the material becomes transparent.

The cutoff wavelength λc calculated for GaAs is 0.87 μm. The excess carriers, compared to the equilibrium state, are created using radiation. Let $\Phi_0(E)$ represent the incident photon flux with coordinates (E,λ), expressed in photons per second per square centimeter (ph/s/cm^2).

If R(E) is the reflection coefficient of the semiconductor for radiation with energy E, the transmitted photon flux $\Phi_t(E)$ at the surface is given by:

$$\varnothing_t(E) = (1 - R(E))\varnothing_0(E) \tag{12}$$

The photons are absorbed during their propagation through the material. The absorption coefficient alphaα is defined as the relative variation in the radiation's density per unit of length. This parameter is provided in **Figure 5** and is calculated analytically using Eq. (13). Thus, if the radiation's density varies by dΦ over a length dx, α is given by:

$$\alpha(E,x) = \frac{-1}{dx}\frac{d\varnothing(E,x)}{\varnothing(E,x)} \tag{13}$$

α represents the relative change in the radiation's density per unit length as it propagates through a material. Mathematically, it is defined as: α = − (dΦ/dx) /Φ where Φ is the radiation's density and x is the distance traveled.

5.1 Now, discussing α and its sign

α is positive when dΦ is negative, i.e., when there is absorption of radiation. If α is constant throughout the material, i.e., α(E,x) = α(E), then integrating the previous expression gives:

$$\frac{d\varnothing(E,x)}{\varnothing(E,x)} = -\alpha(E)dxLn(\varnothing(E,x)) = Ln(\varnothing_t(E)) - \alpha(E)x$$
$$\varnothing(E,x) = \varphi_t(E)e^{-\alpha(E)x} \tag{14}$$

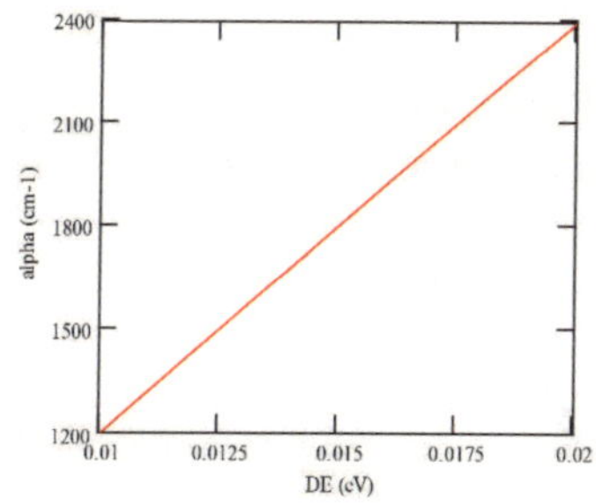

Figure 5.
The absorption coefficient of GaAs versus the energy difference.

- α is positive when there is absorption of radiation. This is because:
- $d\Phi$ is negative when the radiation density decreases (due to absorption).
- A negative $d\Phi$ results in a positive α according to the definition.

Given the complexity of quantum models, phenomenological formulas derived from experience are often recommended. Simple analytical expressions that describe physical phenomena are reported and used [37].

5.2 Absorption coefficient model for GaAs

A specific model for GaAs relates the absorption coefficient to the photon energy and the bandgap energy (E_g) [37].

$$\alpha(h\upsilon) = k\left(h\upsilon - E_g\right)^{1/2} \tag{15}$$

Where:

hυ is the photon energy and Eg is the band gap energy.

Example:

For GaAs, k = 12,000 $cm^{-1}\ eV^{-1/2}$ and for a photon whose energy exceeds 0.01 eV of the band gap energy (E_g), the calculated absorption coefficient is 1200 cm^{-1}.

The calculated absorption coefficient is presented in (**Figure 5**).

This coefficient depends on the material used and the wavelength. However, there are other properties characterizing the photodetector, such as:

- Quantum efficiency (η)
- Sensitivity (S)
- Photocurrent (Iph)
- Optical power received (Popt) ...

6. Principle of the photodetection in the PIN photodiode

Photons transmitted through the optical fiber penetrate the detector, which is composed of semiconductor material. Upon absorption, these photons can cause electrons to transition from the valence band to the conduction band. In the conduction band, electrons become less bound and free; thus, the photon gives rise to an electron-hole pair. An external potential is applied to prevent the electron from returning to its stable state in the valence band. Under the influence of the electric field, the carriers (electrons and holes) are separated and move toward regions where they are majority carriers. These generated carriers are then collected as a photocurrent. The number of electron-hole pairs created is proportional to the number of absorbed photons.

To ensure efficient photodetection and prevent recombination of the electron-hole pairs, absorption must occur in a depleted region. This highlights the advantage of using a PIN photodiode over a standard PN structure [35, 38]. The structure of the PIN photodiode is shown in **Figure 6**.

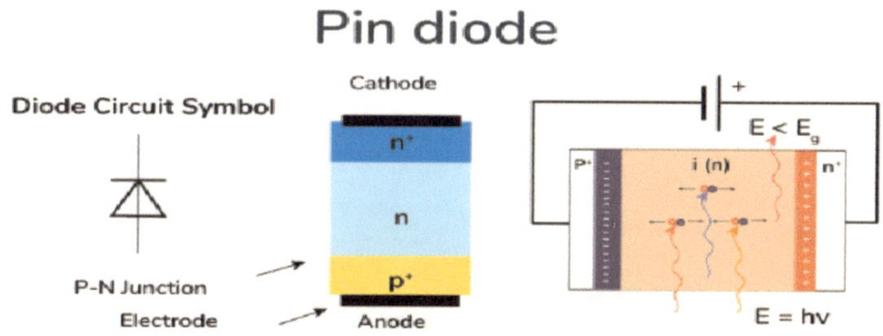

Figure 6.
Photodiode PIN structure.

Most photons are absorbed in the intrinsic region (I), where a nearly uniform electric field exists to separate the carriers. One of the regions traversed by light (either P or N) must be very thin and coated with an anti-reflective layer to enhance external quantum efficiency and protect the material.

Key parameters that characterize the PIN photodiode, depending on the material and structure, include sensitivity, dark current, and response time. Sensitivity depends on the quantum efficiency, which varies with the material's absorption coefficient and the width of the intrinsic region. Dark current can originate from various sources, including thermal generation in the intrinsic region, surface currents, and leakage currents.

The expression for the total current in the photodiode is given by Eq. (16).

$$I_{ph} = S.P_{opt} + I_{dark} \tag{16}$$

The response time of the photodiode must be very short. It is limited by two factors: the junction capacitance (a factor calculated in the following sections) and the transit time. The junction capacitance is reduced when the absorption zone is widened. On the other hand, the transit time decreases when this zone is narrowed. Therefore, a compromise must be found.

The electrical signal emitted by the photodiode is not perfect but is disturbed. In micro-optoelectronics, mobility is a key factor of noise [36]. Other sources of noise in telecommunications are discussed [37], such as background noise (also known as photonic noise or Schottky noise). This noise originates from the randomness in the creation of electron-hole pairs. It is proportional to the optical power received by the photodiode.

7. Simulation steps of conduction in a PIN structure

The characteristics of the structure without illumination are used in the simulation under illumination, allowing for a simplified approach.

7.1 Basic equations involved in the physical model

Solving such a system, in the most general case of a real component with three-dimensional (3D) geometry, while considering boundary and initial conditions, is usually very complex. Therefore, additional simplifying assumptions must be made. The systems of equations that describe the transport mechanisms in a PIN diode at steady state, which will be solved later using numerical simulation, are presented as follows [6, 10, 26–28].

$$\text{div}\left(\varepsilon\ .\overrightarrow{\text{grad}}\ \psi\right) = e\left[n - p + N_A + n_r - p_t\right] \tag{17}$$

$$\begin{aligned} \frac{1}{e}.div\ \overrightarrow{J_n} &= U_r - U_t \\ \frac{1}{e}.\text{div}\ \overrightarrow{J_p} &= -U_r - U_t \end{aligned} \tag{18}$$

with:

$$\begin{aligned} \overrightarrow{J_n} &= e.\mu_n.n.\xi + e.D_n.\overrightarrow{\text{grad}}\,n \\ \overrightarrow{J_p} &= e.\mu_p.p.\xi - e.D_p.\overrightarrow{\text{grad}}\,p \\ \xi &= -\overrightarrow{\text{grad}}\,\psi;\ D = \frac{\mu.k\,T}{e}. \end{aligned} \tag{19}$$

and

$$\begin{aligned} U_{nr} = U_{pr} = U_r &= \frac{n.p - n_i^2}{\tau_{pre}.(n + n_{1r}) + \tau_{nre}.(p + p_{1r})} \\ U_{nt} = U_{pt} = U_t &= \frac{n.p - n_i^2}{\tau_{pte}.(n + n_{lt}) + \tau_{nte}.(p + p_{lt})} \end{aligned} \tag{20}$$

Similarly, the system of equations governing the conduction of minority carriers in the volume of the PIN photodiode is presented below. When the semiconductor is photo-excited, the surplus carriers generated diffuse and recombine. The behavior of this system at steady state and for a three-dimensional analysis is governed by the fundamental equations (Poisson and continuity), as follows [6, 10]:

$$\begin{aligned} &D_p.\left(\frac{\partial^2 P_n}{\partial x^2} + \frac{\partial^2 P_n}{\partial y^2} + \frac{\partial^2 P_n}{\partial z^2}\right) + G(\lambda) - U_r(\lambda) = 0 \\ &D_n.\left(\frac{\partial^2 n_p}{\partial x^2} + \frac{\partial^2 n_p}{\partial y^2} + \frac{\partial^2 n_p}{\partial z^2}\right) - G(\lambda) + U_r(\lambda) = 0 \\ &G(\lambda) = \alpha(\lambda)F(\lambda)(1 - R(\lambda))\exp(-\alpha(\lambda)x) \end{aligned} \tag{21}$$

Where $G(\lambda)$ is the generation rate, α and R are the absorption and reflection coefficients, respectively. F is the number of incident photons per cm^2/s per unit of bandwidth, and λ is the wavelength of the incident light, n_p and p_n, are the minority carrier densities in the P and N regions, respectively. The resulting photocurrent is the sum of three components: the diffusion current of photoelectrons in the P-region, the photogenerated current in the space charge region, and the diffusion current of photoholes in the N-region. The total photocurrent is calculated at the same point, for instance, at the point x_n [6, 10, 26–28]:

$$\begin{aligned} &J_{ndiff}(\lambda) = qD_n\left(\frac{\partial n_p}{\partial x} + \frac{\partial n_p}{\partial y} + \frac{\partial n_p}{\partial z}\right)_{x_j+w} \\ &J_{pdiff}(\lambda) = -qD_n\left(\frac{\partial p_n}{\partial x} + \frac{\partial p_n}{\partial y} + \frac{\partial p_n}{\partial z}\right)_{x_j} \\ &J_{ph} = J_{ndiff}(x_n, y, z) + J_g(x_n, y, z) + J_{pdiff}(x_n, y, z) \end{aligned} \tag{22}$$

7.2 Assumptions and simplifications

We investigated an important characteristic found in a previous study [39], which revealed that the influence of hole trap levels in GaAs (**Table 1**) on conduction is significant near the first side of the diode, while negligible effects are observed in the bulk region. This observation leads us to focus meshing and discretization around the first contact. The principle of current density conservation is applied. Recombination in the space charge region is neglected, and it is assumed that all electrons arriving at x_p from the P-region reach x_n. Therefore, the electron diffusion current at $x = x_n$ is identical to that at $x = x_p$, resulting in [6, 10]:

$$J_{ndiff}(x_n, y, z) = J_{ndiff}(x_p, y, z) \tag{23}$$

Now consider the generation current. At an unspecified point within the space charge region, this current results from both electrons and holes [6, 10].

$$J_g(x, y, z) = J_{gn}(x, y, z) + J_{gp}(x, y, z) \tag{24}$$

At $x = x_p$, the generation current is solely due to holes arriving at x_p, created throughout the space charge zone from x_n to x_p:

$$J_g(x_p, y, z) = J_{gp}(x_p, y, z) \tag{25}$$

On the other hand, at $x = x_n$, the generation current is due only to the electrons created from x_p to x_n [6, 10, 26–28]:

$$\begin{aligned} J_g(x_n, y, z) &= J_{gn}(x_n, y, z) \\ J_{ph} &= J_{ndiff}(x_p, y, z) + J_{gn}(x_n, y, z) + J_{pdiff}(x_n, y, z) \end{aligned} \tag{26}$$

The generation current in the space charge region is described by the following expression [6, 10]:

$$J_g(\lambda) = q\int_{x_j}^{x_j+w} G(\lambda)d\lambda = qF(1-R)\,exp\,(-\alpha x_j)(1-\,exp\,(-\alpha w)) \tag{27}$$

Considering the dual origin of the photocurrent, the response time of the photodiode is determined by the diffusion of the minority photo-carriers from the neutral regions to the space charge region. The first phenomenon is relatively slow and results in a time constant of about 10^{-8}–10^{-9} s. The second phenomenon can be very fast if the reverse bias of the diode is significant. In this case, the carriers cross the space charge region (w) at their saturation velocity (Vsat), and the transit time is given by w/Vsat, which is on the order of 10^{-10}–10^{-11} s.

To minimize the diode's time constant, the radiation should be primarily absorbed in the space charge region, where the generation current is produced. Therefore, a photodiode is designed with the thinnest possible front zone and a sufficiently thick space charge region (w) to absorb most of the incident radiation. Considering these constraints, the chosen value for w does not exceed the ratio w/Vsat, and the selected w varies approximately as $\sim 1/\alpha$. The PIN diode used in the calculations satisfies this condition. The polarization of the structure is sufficiently high, leading to a strong

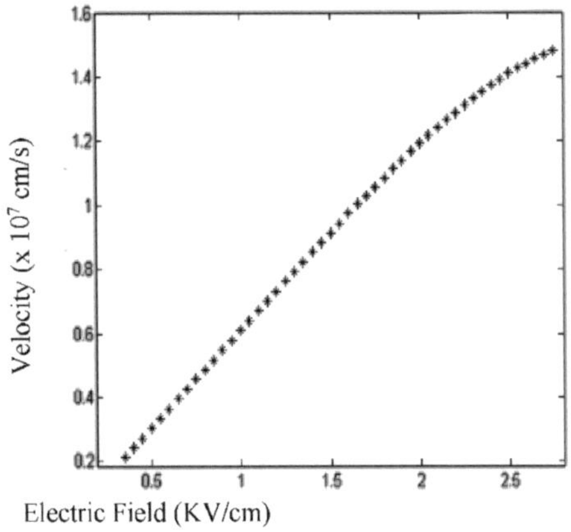

Figure 7.
The velocity variation versus the Electric Field at a reverse bias of 5 volts.

electric field across the intrinsic region, which enables photo-carriers to rapidly reach their saturation velocity, Vsat (see **Figures 7** and **8**). As a result, the photodiode exhibits both high speed and high sensitivity. The electric field arises from two contributions:

$$\xi = \xi_a + \xi_i \tag{28}$$

Here, ξ_a represents the electric field due to polarization, while ξ_i represents the internal field resulting from the difference in mobility between electrons and holes.

The following assumptions are made:

$$\begin{aligned} &\Delta_n(x, y, z) \sim \Delta_p(x, y, z) \\ &\tau_n = \tau_p = \tau \\ &n = n_0 + \Delta_n \\ &p = p_0 + \Delta_p \end{aligned} \tag{29}$$

Given the substantial difference between the dielectric relaxation time 10^{-12} s and the carrier lifetime of 10^{-9}–10^{-3} s, the electrical neutrality of the material is restored on a timescale significantly shorter than the carrier relaxation time. This indicates that the dielectric response of the material occurs much faster than the time required for charge carriers to recombine or become trapped. The excess electrons and holes generated by photoexcitation diffuse together through ambipolar diffusion. Furthermore, since carriers recombine in pairs, the recombination rates of electrons and holes are inherently equal, leading to identical carrier lifetimes:

$$\tau_n = \tau_p = \tau \tag{30}$$

7.3 Boundary conditions

Two types of boundary conditions are implemented in this analysis: Dirichlet conditions and standard Neumann conditions. The following assumptions are made [6]:

- Perfect ohmic contacts.
- A long intrinsic (ν)-region as in our case.

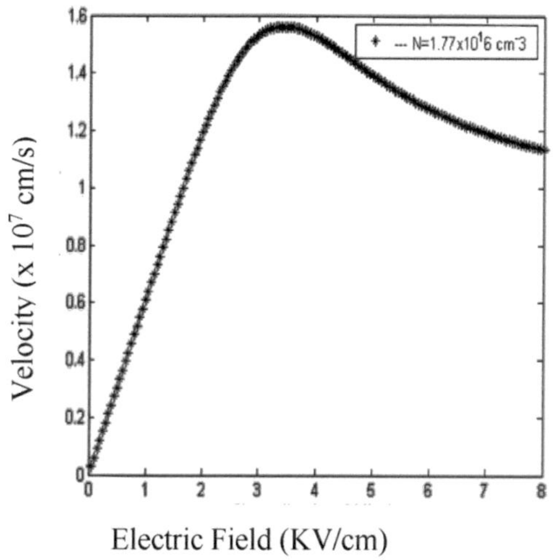

Figure 8.
The velocity variation versus the Electric Field at a reverse bias of 10 volts.

- The electric field is assumed to exist only within the depletion region.
- The drift current is generated exclusively within the depletion region.

The concepts of equilibrium equations and minority carrier continuity are applied.

1. At the surface where x = 0:

$$\Psi = 0; \mathrm{P_n} = \mathrm{P_{n0}} \tag{31}$$

Where P_{n0} is the density of the minority carrier (hole) when the N-type semiconductor is in equilibrium.

2. At the edge of the depletion region, the position x is denoted as x_p, which, under our assumptions, coincides with the contact boundary between the P^+ and ν regions:

$$\begin{aligned} \psi &= Va + \frac{\mathrm{kT}}{q} ln\left(\frac{n_i}{p}\right) \mathrm{Va : is\ the\ reverse\ bias} \\ n_p &= n_{p0}\ e^{\left(\frac{qVa}{kT}\right)} \end{aligned} \tag{32}$$

Photo current generated by the P-type (J_p) semiconductor can be accounted for. Note that the carriers considered here are electrons.

$$J_{ndiff} = J_p \ = qD_n\left(\frac{\partial n_p}{\partial x} + \frac{\partial n_p}{\partial y} + \frac{\partial n_p}{\partial z}\right) \tag{33}$$

Similarly, photo current generated by the N-type at the edge where x = x_p can be accounted for. Note that the carriers considered here are holes.

$$\mathrm{P_n} = \mathrm{P_{n0}}\ e^{\left(\frac{qVa}{kT}\right)} \tag{34}$$

3. At the other boundaries, Neumann conditions are applied, where the normal components of the hole and electron current densities are set to zero, indicating that the normal derivatives of the quantities ψ, n, and p are set to zero.

8. Description of the simulator

The numerical simulation software is written in C++ version 5, under Windows, which enables the declaration of large three-dimensional arrays and improves memory management. It implements an iterative process that combines Newton's linearization of the system, Successive Over-Relaxation (SOR) for solving the linearized systems, and a multigrid approach to efficiently correct residuals at different grid scales [21, 34, 39–49]. The solution data are plotted using MATLAB.

The simulator enables the simulation of the conduction in components based on relaxation semiconductors. It considers the effects of deep centers in relaxation semiconductor devices. The matrix dimensions used are [90] × [90] × [90].

The principal subroutines and steps that form the simulator are illustrated in the schematic diagrams (**Figure 9**).

The software includes the following steps.

1. Linearization via Newton's method:

   The coupled system of Poisson and continuity equations is linearized using Newton's method, which involves approximating the system around an initial guess.

2. *SOR relaxation on the linearized system:*

   - Initialize unknown variables close to their solution.
   - Solve the system using a discretized expression.
   - Perform Successive Over-Relaxation (SOR) sweeps on the system.
   - A convergence test is applied to check the residuals, and if convergence is achieved, update the variables while considering boundary conditions.
   - The applied voltage is gradually increased in steps.

3. *Weighted residual transfer to coarser levels and SOR on Level 1:*

   - Transfer the residuals from the fine grid to coarser grids (Level 2 and Level 1) with weighted corrections.
   - Perform N SOR iterations on Level 1 to correct the errors.

4. *Prolongation to Level 2 and SOR iterations:*

   - Prolong the results obtained on Level 1 to Level 2 (a finer grid).
   - Perform M SOR iterations to refine the solution on this level.

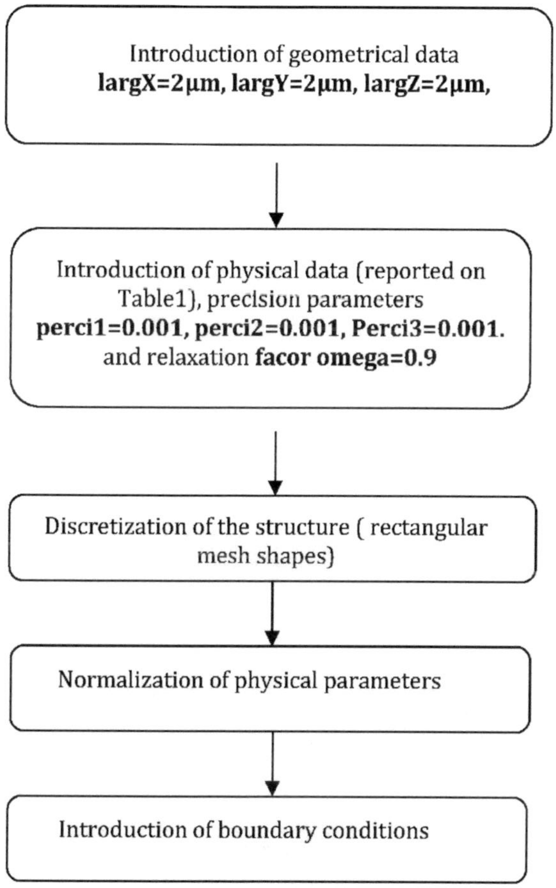

Figure 9.
Introduction data flow chart.

8.1 Diagram schemes description

The flow chart of **Figure 9** illustrates the following key processes:

1. Introduction of geometrical and physical data as reported in the diagram and in **Table** 1.

2. Discretization of the structure.

3. Normalization of parameters.

4. Application of boundary conditions.

The structure is discretized using using a rectangular mesh of 20x20x20 nodes is employed [40, 50]. This involves dividing the simulation domain into smaller elements to facilitate numerical analysis.

Physical parameters are normalized to standardize units and ensure numerical stability during the simulation. The carrier densities n and p typically vary over a wide range (from 10^{-3} to 10^{23} cm^{-3}), which makes it challenging to obtain an accurate evaluation of these quantities in numerical calculations. Therefore, a suitable change of variables (normalization) is applied to enhance the precision of the calculations.

- *Dirichlet conditions*: Specify fixed values at the boundaries.

- *Neumann conditions*: Specify the derivative of values at the boundaries, ensuring proper representation of physical conditions at the edges of the simulation domain.

The algorithm scheme is illustrated below **Figure 10**.

- The structure under investigation is rectangular, and the suitable technique for discretization in this case is the finite difference method [40, 50].

To summarize, the system of equations (poisson and continuity)is discretized by first transforming the differential equations into linear equations (Eqs. (17), (19), (21), and (22)). These are then discretized on a three-dimensional grid with either a uniform or variable mesh size (using a geometric series) through the finite difference method. After assembling the discretized equations, a matrix equation is formulated. To solve this matrix equation, the Gummel and Newton methods [41–49] are employed.

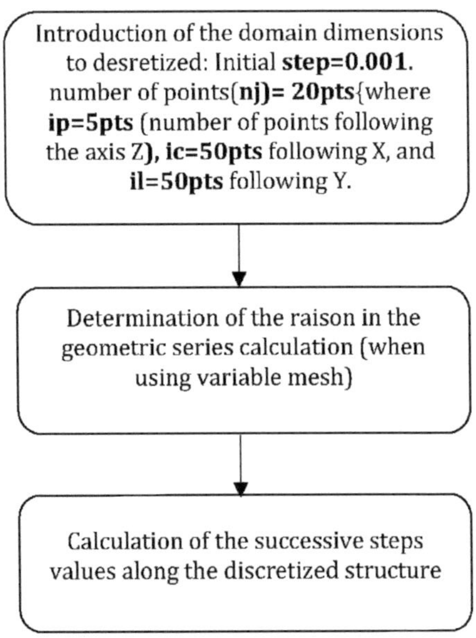

Figure 10.
Discretization flow chart.

The multigrid approach is combined with the Newton algorithm to accelerate convergence across different grid levels. The resolution is three-dimensional, meaning that each node in the discretization has six nearest neighbors (two in each direction). Therefore, the Jacobian matrix is a band matrix with seven nonzero diagonal elements. The system to be solved consists of three equations for each mesh point on the grid.

9. Computational algorithm

As described earlier, the characteristics used in the simulation include the W-cycle with full weighting; assume that three grid levels are used (being the finest grid). The simulator is based on the theorem [41–49]:

1. Linearization of the coupled system,

2. M SOR relaxation by:

   - Initializing the values of unknowns Ψ, n and p (to values close to the solution).
   - Solving the system of N equations for N unknowns using a simple discretized expression. At each point in the discretization. The physical quantity calculated is determined by the contribution of the point itself, along with those of its six nearest neighbors in the X, Y, and Z directions.

In this step, we calculate the full Jacobian matrix U and the vector $\vec{F}$ representing the discretized equations to solve the system $U\delta = -F$, where δ is the correction vector. Each element of the vector is a three-component vector, representing the Poisson and continuity equations:

$$\left(-F_{\psi}, -F_{n}, -F_{P}\right) \tag{35}$$

The convergence test evaluates the number of erroneous points. Once the convergence of the linear system is achieved, the variables are updated while taking the boundary conditions into account. The applied voltage is then increased in increments.

3. Weighted transfer of residuals to levels 2 and 1, followed by M SOR iterations at level 1.

4. Prolongation of the result to level 2, followed by M SOR iterations.

5. Prolongation of the result and update of the linearized system at level 3.

6. Update of the Newton step at level 3.

One multigrid iteration includes steps 3, 4, and 5. At the final stage of the algorithm, the electric field and current density are calculated.

As part of the W-cycle, steps 3 and 4 are repeated twice. The optimized number of multigrid iterations per Newton step is 3.

We have also implemented damping criteria that monitor the corrections made to the variables to ensure they remain compatible with the stability of the resolution process.

This prevents divergence and ensures that the solution stays within a range of physically acceptable values.

The numerical solution in the Newton method is stopped once a convergence criterion is met. The stopping test used is defined by:

$$\left|\frac{\delta\Psi}{\Psi}\right| \leq \varepsilon \quad \left|\frac{\delta n}{n}\right| \leq \varepsilon \quad \left|\frac{\delta p}{p}\right| \leq \varepsilon \tag{36}$$

Where is a predefined value at all mesh points. It provides an indication of the convergence of the resolution method [51–53]. There is another fundamental criterion to assess the quality of the solution, which is the verification of the principle of the conservation of total current.

At a given discretization point, it is not necessary for all the diagonal elements of the Jacobian matrix to contribute to the calculation. This avoids numerical overflow and underflow problems.

9.1 Model limitations

A numerical scheme that solves the continuity equations should [39]:

- Conserve the total number of particles within the simulated device.
- Ensure the local positivity of carrier density, as negative density is unphysical.
- Maintain the monotonicity of the solution, meaning it should not introduce spurious spatial oscillations.

When discretizing the equations, in our case, using finite differences, there are limitations on the choice of mesh size and time step:

- The mesh size is constrained by the Debye length L_{Da}.
- The time step is limited by the dielectric relaxation time τ.

10. Numerical resolution

We present a discretization scheme for the system of equations governing conduction in obscurity, where the equations are more general and include drift-diffusion terms. Discretizing the continuity equations in conservation form requires determining the currents at the midpoints of the mesh lines connecting neighboring grid nodes. Since solutions are available only at the grid nodes, interpolation schemes are necessary to calculate the currents at these midpoints.

To ensure consistency with Poisson's equation, it is common to assume that the potential varies linearly between two neighboring nodes.

This assumption leads to the concept of a constant electric field along the mesh lines, and the field at the midpoint is obtained using centered finite differences of the potential values. To evaluate the current density, it is also necessary to estimate the carrier density at the midpoints. The simplest approximation is to assume a linear variation of the carrier density, which involves taking the arithmetic average of the densities at two neighboring nodes. However, this approximation is only valid for very small potential variations between nodes and is exact only if the electric field between the two nodes is zero, meaning that the carrier densities at both points are identical.

To illustrate this, consider a 3-D mesh where we aim to discretize the electrostatic potential. The Poisson equation to be discretized at the grid points (i, j, l) [21–23, 25, 54] is:

$$\frac{\partial^2\psi}{\partial X^2}+\frac{\partial^2\psi}{\partial Y^2}+\frac{\partial^2\psi}{\partial Z^2}=N_{i,j,l}-P_{i,j,l}-(N_D-N_A)_{i,j,l} \tag{37}$$

The variable ψ is assumed to vary linearly between two successive points. The left side of Poisson's equation is expressed using a centered difference scheme:

$$\frac{\partial}{\partial X}\left(\frac{\partial\psi}{\partial X}\right)\bigg|_i+\frac{\partial}{\partial Y}\left(\frac{\partial\psi}{\partial Y}\right)\bigg|_j+\frac{\partial}{\partial Z}\left(\frac{\partial\psi}{\partial Z}\right)\bigg|_l=\frac{\frac{\partial\psi}{\partial X}\Big|_{i+\frac{1}{2},j,l}-\frac{\partial\psi}{\partial X}\Big|_{i-\frac{1}{2},j,l}}{X_{i+\frac{1}{2},j,l}-X_{i-\frac{1}{2},j,l}}+\frac{\frac{\partial\psi}{\partial Y}\Big|_{i,j+\frac{1}{2},l}-\frac{\partial\psi}{\partial Y}\Big|_{i,j-\frac{1}{2},l}}{Y_{i,j+\frac{1}{2},l}-Y_{i,j-\frac{1}{2},l}}$$
$$+\frac{\frac{\partial\psi}{\partial X}\Big|_{i+1/2,j,l}-\frac{\partial\psi}{\partial X}\Big|_{i,j,l-1/2}}{Z_{i,j,l+1/2}-Z_{i,j,l-1/2}}=N_{i,j,l}-P_{i,j,l}-(N_D-N_A)_{i,j,l} \tag{38}$$

Where:

$$\begin{aligned}X_{i+1/2,j,l}&=X_{i,j,l}+\frac{h_i}{2};X_{i-1/2,j,l}=X_{i,j,l}-\frac{h_{i-1}}{2}\\Y_{i,j+1/2,l}&=Y_{i,j,l}+\frac{kj}{2};Y_{i,j-1/2,l}=Y_{i,j,l}-\frac{k_{j-1}}{2}\\Z_{i,j,l+1/2}&=Z_{i,j,l}+\frac{l_l}{2};Z_{i,j,l-1/2}=Z_{i,j,l}-\frac{l_{l-1}}{2}\end{aligned} \tag{39}$$

with

$$\begin{aligned}\frac{\partial\psi}{\partial X}\bigg|_{i+1/2,j,l}&=\frac{\psi_{i+1,j,l}-\psi_{i,j,l}}{h_i};\frac{\partial\psi}{\partial X}\bigg|_{i-1/2,j,l}=\frac{\psi_{i,j,l}-\psi_{i-1,j,l}}{h_{i-1}}\\\frac{\partial\psi}{\partial Y}\bigg|_{i,j+1/2,l}&=\frac{\psi_{i,j+1,l}-\psi_{i,j,l}}{k_j};\frac{\partial\psi}{\partial Y}\bigg|_{i,j-1/2,l}=\frac{\psi_{i,j,l}-\psi_{i,j-1,l}}{k_{j-1}}\\\frac{\partial\psi}{\partial Z}\bigg|_{i,j,l+1/2}&=\frac{\psi_{i,j,l+1}-\psi_{i,j,l}}{l_l};\frac{\partial\psi}{\partial Z}\bigg|_{i,j,l-1/2}=\frac{\psi_{i,j,l}-\psi_{i,j,l-1}}{l_{l-1}}\end{aligned} \tag{40}$$

The right-hand side of Poisson's equation represents the space charge density, which is expressed in terms of the values of the unknowns at (i,j,l).

$$\left(-\rho_{i,j,l}\right)=N_{i,j,l}-P_{i,j,l}-(N_D-N_A)_{i,j,l} \tag{41}$$

Where ρ denotes the charge density, we obtain a Poisson equation that is linear in ψ, discretized as follows:

$$F_{\psi}^{i,j,l} = A_{i-1,j,l}.\left(\psi_{i-1,j,l} - \psi_{i,j,l}\right) + A_{i+1,j,l}.\left(\psi_{i+1,j,l} - \psi_{i,j,l}\right) + A_{i,j-1,l}.\left(\psi_{i,j-1,l} - \psi_{i,j,l}\right)$$
$$+A_{i,j+1,l}.\left(\psi_{i,j+1,l} - \psi_{i,j,l}\right) + A_{i,j,l-1}\cdot\left(\psi_{i,j,l-1} - \psi_{i,j,l}\right) + A_{i,j,l+1}\cdot\left(\psi_{i,j,l+1} - \psi_{i,j,l}\right) + \rho_{i,j,l} = 0 \quad (42)$$

Due to the presence of exponential terms in the expressions for N and P concentrations, Poisson's equation must be linearized [26]. This linearization is achieved using a Taylor series expansion, as follows:

$$\Psi = \Psi^0 + \delta\Psi \quad (43)$$

Where $\delta\Psi$ is a very small increment compared to Ψ.

For the electron and hole densities, we write:

$$N(\Psi) = N\left(\Psi^0\right) + \left.\frac{\delta N(\Psi)}{\delta\Psi}\right|_{\Psi-\Psi^0}.\delta\Psi$$
$$\mathrm{P}(\Psi) = \mathrm{P}\left(\Psi^0\right) + \frac{\delta P(\Psi)}{\delta\Psi}\left|\frac{\delta P(\Psi)}{\delta\Psi}\right|_{\Psi-\Psi^0}.\delta\Psi \quad (44)$$

The discretized expression of the Poisson equation is then written:

$$-\mathrm{coef}_{i,j,l}.\delta\Psi_{i,j,l} + A_{i+1,j,l}.\delta\Psi_{i+1,j,l} + A_{i-1,j,l}.\delta\Psi_{i-1,j,l} + A_{i,j+1,l}.\delta\Psi_{i,j+1,l} + A_{i,j-1,l}.\delta\Psi_{i,j-1,l}$$
$$+A_{i,j,l+1}.\delta\Psi_{i,j,l+1} + A_{i,j,l-1}.\delta\Psi_{i,j,l-1} = b_{i,j,l} \quad (45)$$

The discretized form of the electron three-dimensional continuity equation is written as:

$$\frac{\partial J_n}{\partial X} + \frac{\partial J_n}{\partial Y} + \frac{\partial J_n}{\partial Z} = U_{i,j,l}$$
$$U_{i,j,l} = \frac{N_{i,j,l,.P_{i,j,l-1}}}{\tau_{nr}\left(P_{i,j,l} + P_1\right) + \tau_{Pr}\left(N_{i,j,l} + N_1\right)}\text{(after normalization)} \quad (46)$$

The discretized continuity equations for electrons and holes are given in the following form [28, 35, 39]:

$$F_n^{i,j,1} = \mu_n\begin{bmatrix}\alpha_{i-1,j,1}.N_{i-1,j,1} - \left(\alpha_{i,j,1} + \alpha_{i-1,j,1}.\,\mathbf{exp}\left(\Psi_{i-1,j,1} - \Psi_{i,j,l}\right)\right)N_{i,j,1} \\ +\alpha_{i,j,1}.\,\mathbf{exp}\left(\Psi_{i,j,1} - \Psi_{i+1,j,l}\right)N_{i+1,j,l}\end{bmatrix}\cdot\frac{1}{x_{i+1,j,1} - x_{i-1,j,l}}$$
$$+\mu_n\left[\beta_{i,j-1,l}.N_{i,j-1,1} - \left(\beta_{i,j,l} + \beta_{i,j-1,l}.\,\mathbf{exp}\left(\Psi_{i,j-1,1} - \Psi_{i,j,1}\right)\right)N_{i,j,1}\right.$$
$$\left.+\beta_{i,j,1}.\,\mathbf{exp}\left(\Psi_{i,j,1} - \Psi_{i,j+1,1}\right)N_{i,j+1,1}\right]\cdot\frac{1}{Y_{i,j+1,1} - Y_{i,j-1,1}}$$
$$+\mu_n\left[\gamma_{i,j,l-1}.N_{i,j,l-1} - \left(\gamma_{i,j,l} + \gamma_{i,j,l-1}.\,\mathbf{exp}\left(\Psi_{i,j,l-1} - \Psi_{i,j,l}\right)\right)N_{i,j,l}\right.$$
$$\left.+\gamma_{i,1,1}.\,\mathbf{exp}\left(\Psi_{i,j,l} - \Psi_{i,j,l+1}\right)N_{i,j,l+1}\right]\cdot\frac{1}{Z_{i,i,l+1} - Z_{i,i,l}}$$
$$-\frac{N_{i,j,l}.P_{i,j,l} - 1}{\tau_{n_r}.\left(P_{i,j,l} + P_1\right) + \tau_{p_r}.\left(N_{i,j,l} + N_1\right)} = 0 \quad (47)$$

$$F_n^{i,j,1} = \mu_n \begin{bmatrix} \alpha_{i-1,j,1}.N_{i-1,j,1} - \left(\alpha_{i,j,1} + \alpha_{i-1,j,1}.\,\mathbf{exp}\left(\Psi_{i-1,j,1} - \Psi_{i,j,l}\right)\right)N_{i,j,1} \\ +\alpha_{i,j,1}.\,\mathbf{exp}\left(\Psi_{i,j,1} - \Psi_{i+1,j,l}\right)N_{i+1,j,l} \end{bmatrix} \cdot \frac{1}{x_{i+1,j,1} - x_{i-1,j,l}}$$
$$+\mu_n\left[\beta_{i,j-1,l}.N_{i,j-1,1} - \left(\beta_{i,j,l} + \beta_{i,j-1,l}.\,\mathbf{exp}\left(\Psi_{i,j-1,1} - \Psi_{i,j,1}\right)\right)N_{i,j,1}\right.$$
$$\left.+\beta_{i,j,1}.\,\mathbf{exp}\left(\Psi_{i,j,1} - \Psi_{i,j+1,1}\right)N_{i,j+1,1}\right] \cdot \frac{1}{Y_{i,j+1,1} - Y_{i,j-1,1}}$$
$$+\mu_n\left[\gamma_{i,j,l-1}.N_{i,j,l-1} - \left(\gamma_{i,j,l} + \gamma_{i,j,l-1}.\,\mathbf{exp}\left(\Psi_{i,j,l-1} - \Psi_{i,j,l}\right)\right)N_{i,j,l}\right.$$
$$\left.+\gamma_{i,1,1}.\,\mathbf{exp}\left(\Psi_{i,j,l} - \Psi_{i,j,l+1}\right)N_{i,j,l+1}\right] \cdot \frac{1}{Z_{i,i,l+1} - Z_{i,i,l}}$$
$$-\frac{N_{i,j,l}.P_{i,j,l} - 1}{\tau_{n_r}.\left(P_{i,j,l} + P_1\right) + \tau_{p_r}.\left(N_{i,j,l} + N_1\right)} = 0 \tag{48}$$

After discretization, three systems of equations are obtained, each with 3 N unknowns F_Ψ, Fn, and F_p. The discretized expression for the current density at the midpoint of the mesh line between nodes x_i and x_{i+1} is given by:

$$Jn_{i+1/2} = e.\mu_n \cdot \left[\left(\frac{n_i + n_{i+1}}{2}\right) \cdot \frac{\psi_i - \psi_{i+1}}{x_{i+1} - x_i} + \frac{kT}{e}\frac{n_{i+1} - n_i}{x_{i+1} - x_i}\right] \tag{49}$$

It is possible to deduce the concentrations of electrons and holes by integrating the continuity equations (in terms of the current density) over each elementary mesh interval $[x_i,x_{i+1}]$. For electrons, we obtain:

$$\frac{\partial n}{\partial x} - \frac{e}{kT}\frac{\partial \psi}{\partial x} n = \frac{Jnx}{\mu_n.kT} \tag{50}$$

The integration method known as “the constant’s variation” leads in terms of x to the following solution:

$$n(x) = u(x).\exp\left(\frac{e\,\psi}{kT}\right) \tag{51}$$

with:

$$u(x) = \int \frac{Jnx}{\mu_n.kT}\exp\left(-\frac{e\,\psi}{kT}\right)dx \tag{52}$$

Assuming J_n constant over the interval $[x_i,x_{i+1}]$, we obtain the following equation:

$$Jnx_{i+1/2} = \mu_n.kT.\frac{u_{i+1} - u_i}{\int_{xi}^{xi+1}\exp\left(-\frac{e\,\psi}{kT}\right)dx} \tag{53}$$

with:

$$\begin{aligned} u_{i+1} &= n_{i+1}.\exp\left(-\frac{e\psi_{i+1}}{kT}\right) \\ u_i &= n_i.\exp\left(-\frac{e\psi_i}{kT}\right) \end{aligned} \tag{54}$$

The same reasoning leads, for holes current J_p to the following expression:

$$Jpx_{i+1/2} = \mu_p.kT.\frac{u'_i - u'_{i+1}}{\int_{xi}^{xi+1} exp\left(\frac{e\,\psi}{kT}\right)dx} \tag{55}$$

with:

$$\begin{aligned} u'_{i+1} &= p_{i+1}.\exp\left(\frac{e\,\psi_{i+1}}{kT}\right) \\ u'_i &= p_i.\exp\left(\frac{e\,\psi_i}{kT}\right) \end{aligned} \tag{56}$$

These current formulas (53) and (55) are based on the concentration distributions n(x) and p(x) within each elementary interval $x_{i+}1$-x_i, such that the values of each conduction current Jn and Jp remain constant over this interval. The Gummel expressions for J_n and J_p in the elementary intervals x_{i+1}-x_i, y_{j+1}-y_j and $z_{l+1} - z_l$ are given by:

$$\begin{aligned} Jnx_{i+1/2,j,l} &= e.\mu_n\cdot\frac{n_{i+1,j,l}.exp\left(\frac{e\left(\psi_{i,j,l}-\psi_{i+1,j,l}\right)}{kT}\right)-n_{i,j,l}}{exp\left(\frac{e\left(\psi_{i,j,l}-\psi_{i+1,j,l}\right)}{kT}\right)-1}\cdot\frac{\psi_{i,j,l}-\psi_{i+1,j,l}}{x_{i+1}-x_i} \\ Jny_{i,j+1/2,l} &= e.\mu_n\cdot\frac{n_{i,j+1,l}.exp\left(\frac{e\left(\psi_{i,j,l}-\psi_{i,j+1,l}\right)}{kT}\right)-n_{i,j,l}}{exp\left(\frac{e\left(\psi_{i,j,l}-\psi_{i,j+1,l}\right)}{kT}\right)-1}\cdot\frac{\psi_{i,j,l}-\psi_{i,j+1,l}}{y_{j+1}-y_j} \\ Jnz_{i,j,l+1/2} &= e.\mu_n\cdot\frac{n_{i,j,l+1}\ exp\left(\frac{e\left(\psi_{i,j,l}-\psi_{i,j,l+1}\right)}{kT}\right)-n_{i,j,l}}{exp\left(\frac{e\left(\psi_{i,j,l}-\psi_{i,j,l+1}\right)}{kT}\right)-1}\cdot\frac{\psi_{i,j,l}-\psi_{i,j,l+1}}{z_{l+1}-z_l} \end{aligned} \tag{57}$$

and

$$Jpx_{i+1/2,j,l} = e.\mu_p \cdot \frac{p_{i,j,l} \cdot exp\left(\frac{e\left(\psi_{i,j,l} - \psi_{i+1,j,l}\right)}{kT}\right) - p_{i+1,j,l}}{exp\left(\frac{e\left(\psi_{i,j,l} - \psi_{i+1,j,l}\right)}{kT}\right) - 1} \cdot \frac{\psi_{i,j,l} - \psi_{i+1,j,l}}{x_{i+1} - x_i}$$

$$Jpy_{i,j+1/2,l} = e.\mu_p \cdot \frac{p_{i,j,l} \cdot exp\left(\frac{e\left(\psi_{i,j,l} - \psi_{i,j+1,l}\right)}{kT}\right) - p_{i,j+1,l}}{exp\left(\frac{e\left(\psi_{i,j,l} - \psi_{i,j+1,l}\right)}{kT}\right) - 1} \cdot \frac{\psi_{i,j,l} - \psi_{i,j+1,l}}{y_{j+1} - y_j} \quad (58)$$

$$Jpz_{i,j,l+1/2} = e.\mu_p \cdot \frac{p_{i,j,l} \cdot exp\left(\frac{e\left(\psi_{i,j,l} - \psi_{i,j,l+1}\right)}{kT}\right) - p_{i,j,l+1}}{exp\left(\frac{e\left(\psi_{i,j,l} - \psi_{i,j,l+1}\right)}{kT}\right) - 1} \cdot \frac{\psi_{i,j,l} - \psi_{i,j,l+1}}{z_{l+1} - z_l}$$

- For stability, it is essential to verify that the mesh spacing is sufficiently small. We assume a linear variation of the potential between neighboring mesh points, which is consistent with the use of finite differences to represent the electric field.

- The assumption of potential linearity between mesh points is reflected in the use of centered finite differences to express the field at the midpoint. The field is explicitly determined through linear interpolation while staying within the framework of the Gummel approximation.

- The semiconductor equations form a coupled nonlinear system. A solution cannot typically be obtained in a single step; instead, a multigrid nonlinear iteration method is necessary.

- The natural variable formulation (V, n, p) is employed in the solution process.

- Determining an optimal strategy for the solution is challenging, as it depends on various factors specific to the device under study.

11. Results and comments

The continuity equations are solved using the combined Gummel-Newton method. Notable results are obtained regarding the electric field distribution and the photocurrent profile. The simulation is validated under reverse biases of 5 and 10 V.

The quality of the solution is highly dependent on the mesh used. An inadequate mesh, with insufficient discretization nodes, can lead to significant oscillations or even divergence in some cases.

The analysis of the structure in its thermal equilibrium state provides valuable insight into the potential, free carrier distributions, and the charge state within the photodiode, as illustrated in **Figures 11–15**. This analysis allows for a clear identification of the transport phenomena that emerge after excitation.

The space charge is a function of the potential distribution, i.e., a function of the solutions to the problem. Assuming that the space charge in the charge space zone is empty of free carriers, meaning it contains only fixed charges (the donor and acceptor ions), the spatial distribution of the space charge becomes independent of the potential. Additionally, assuming that the distribution of free carriers n and p matches exactly that of the donors and acceptors in the conducting zones, it is important to take into account the density of trap centers, which influences the contact P^+v.

When the PνN structure (**Figure 4**) is not exposed to external excitation and without illumination, it exhibits certain characteristics (shown in **Figures 11–15**). These characteristics will be modified and disturbed under light.

Due to the large volume of data for the space charge distribution (**Figure 15**), it was difficult to visualize and clearly identify any perturbations or variations between 10^{15} and 10^{17} on the same scale. Therefore, the solution was to split the data into two graphs, with the left graph ending and the right graph beginning at the same point at the bottom of the right graph. This approach allows for easier reading and comparison. The curves, which may have a one-dimensional representation [34], could provide a closer explanation. To better understand how to interpret **Figure 15**, one can compare it with **Figure 16** and observe the similarities between the two.

Based on the curves in **Figures 15** and **16**, we can determine the widths of the space charge regions and analyze them analytically by performing a simple calculation using semiconductor physics equations [6, 26] to validate or not the results. The calculation of the thickness of the space charge region is carried out by considering that it primarily extends into the less-doped area, where the space charge mainly originates from ionized deep-level centers.

$$\rho_{(P^+\nu)} = e \cdot (N_t - N_A) \tag{59}$$

A classical calculation leads to:

$$W_{(P^+-\nu)} = \left(\frac{2.\varepsilon.V_{D(P^+\nu)}}{\rho_{(P^+\nu)}} \right)^{1/2} \tag{60}$$

This gives a space charge region width of 0.243 µm for the $P^+\nu$ contact.

Note that for the calculation of the space charge region width $W_{(\nu\text{-}N^+)}$, the electric charge only comes from the acceptors.

$$\rho_{(\nu\text{-}N^+)} = \text{-}e.N_A \tag{61}$$

$$W_{(\nu-N^+)} = \left(\frac{2.\varepsilon.V_{D(\nu-N^+)}}{\rho_{(\nu-N^+)}} \right)^{1/2} \tag{62}$$

After performing the calculation, we obtain $W_{(\nu\text{-}N^+)}$ = 0.622 µm. In the volume, and considering the model used, the variation in the space charge is associated with a change in the occupancy of the deep-level center EL2, since the acceptors are always fully ionized.

The relationship between the electric field and the applied reverse bias voltage in the photodiode can be seen through **Figure 7** with **Figure 17** and **Figure 8** with **Figure 18** and be explained as follows:

When a reverse bias is applied to the photodiode an electric field is formed at the junction. This electric field opposes the injection of charge carriers (electrons and holes) and creates a depletion region.

The applied reverse bias generates an electric field within this depletion region proportional to the reverse bias voltage. The higher the reverse bias, the more intense the electric field in the depletion region becomes, as in the expression below. This prevents the flow of current, except for a very small leakage current due to minority carriers.

$$\xi = \frac{Vd}{\mathrm{d}} \tag{63}$$

However, when the reverse bias voltage reaches a critical value (called the "breakdown voltage"), the electric field becomes strong enough to cause an avalanche of charge carriers, resulting in a significant reverse current (the diode's breakdown effect). The speed of charge carriers in a diode depends on the applied electric field and the material properties of the semiconductor. This field acts on the charge carriers, and their speed increases with the intensity of the field.

In a moderate electric field (*linear regime*), charge carriers (electrons in the N-region and holes in the P-region) move at a speed that is directly proportional to the applied field and reach saturation velocity. This is described by the relation [6]:

$$Vd = \frac{\mu.\xi}{\sqrt{1 + \left(\frac{\mu.\xi}{Vsat}\right)^2}} \tag{64}$$

where:

- *Vd* is the velocity of the charge carriers,
- μ is the mobility of the charge carriers,
- V_{sat} is the saturation velocity.

However, beyond a certain electric field, the carrier velocity reaches a limiting value, known as the saturation velocity. At this point, even if the electric field continues to increase, the charge carriers cannot accelerate any further. This is due to the repeated collisions of carriers with the atoms of the crystal lattice of the semiconductor, which limit their acceleration.

The saturation velocity is reached when the charge carriers are energetic enough to collide with the crystal lattice at a high frequency, dissipating the additional energy they acquire.

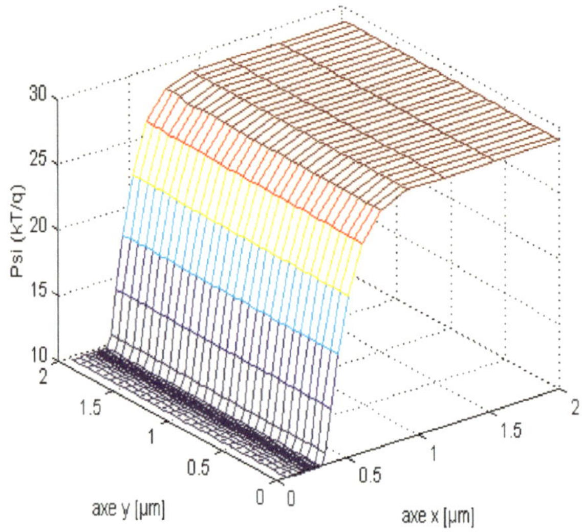

Figure 11.
Profile of electrostatic potential at equilibrium.

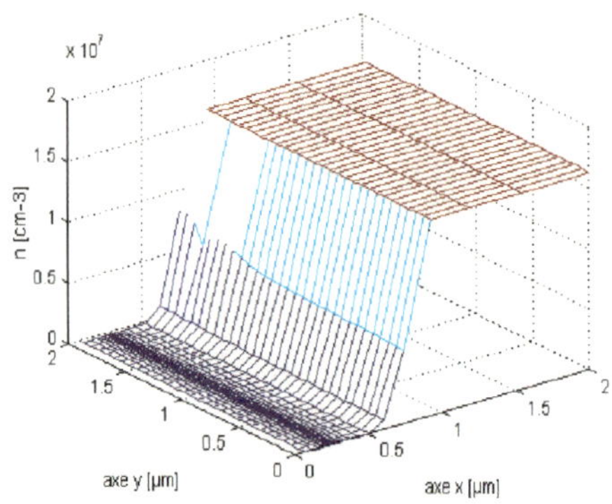

Figure 12.
3D profile of free electron concentration at equilibrium.

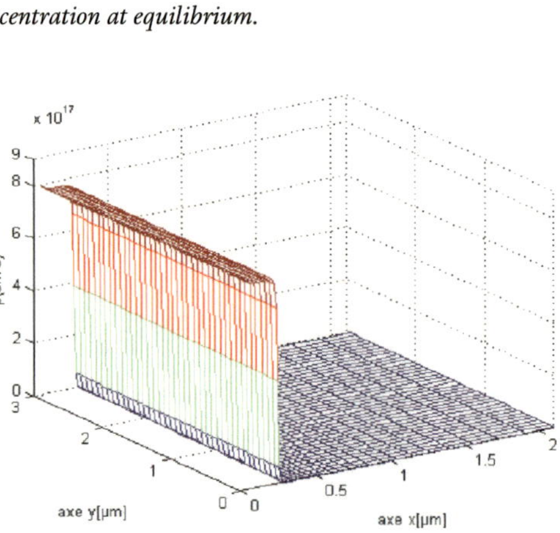

Figure 13.
3D profile of free holes concentration at equilibrium.

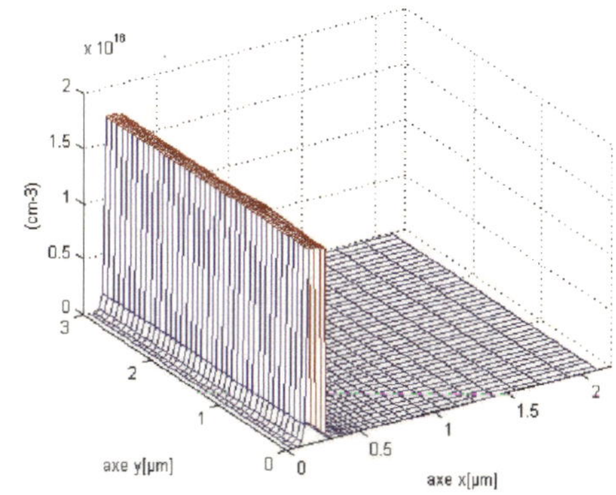

Figure 14.
3D Profile of space charge at equilibrium state in $P^+\nu$ contact.

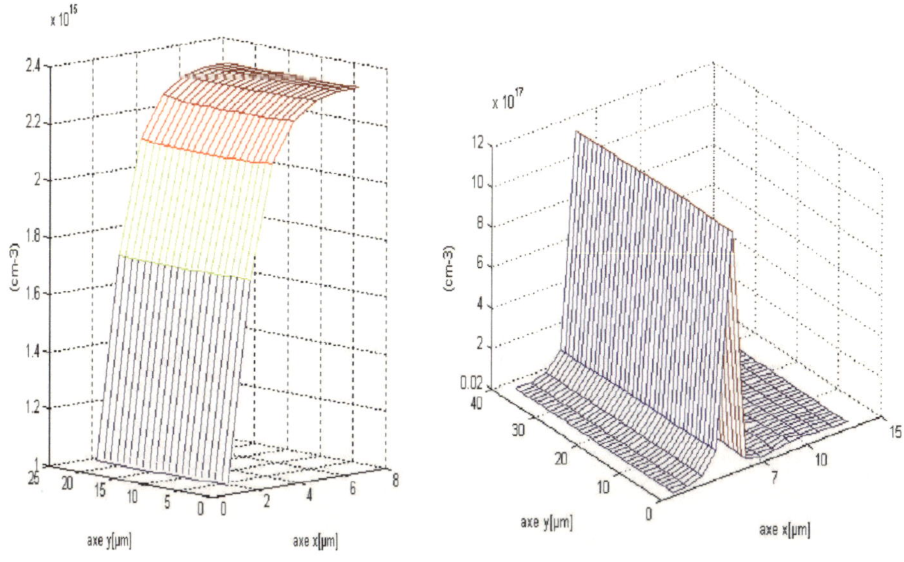

Figure 15.
3D Profile of space charge at equilibrium state in νN^+ contact.

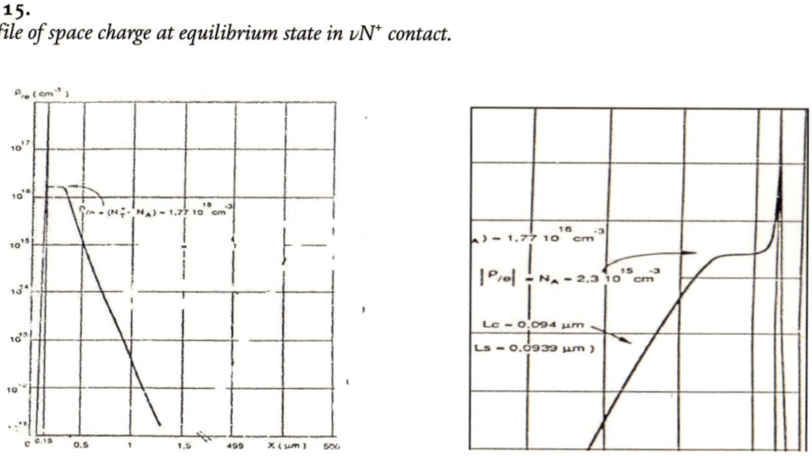

Figure 16.
Space charge profile in the $P^+\nu$ and νN^+ contact at thermodynamic equilibrium - one-dimensional simulation [34].

Now, if the reverse bias voltage increases sufficiently, the electric field becomes stronger, and the carriers accelerate, potentially leading to an avalanche phenomenon. In this case, additional carriers are generated by collisions, exponentially increasing the reverse current. Because the high values of the electric field (**Figure 18**), 10 volts can be seen as high level injection in GaAs. The adopted expression of velocity versus electric field is as follows [20, 55]:

$$Vd = \frac{\mu_0 E + V_{sat}(E/E_c)^4}{1 + (E/E_c)^4} \tag{65}$$

E_c is the critical value of the electric field.
μ_0 is the mobility at weak field given by [56]

$$\mu 0 = \frac{7200}{1 + \left[\frac{n}{nref}\right]^{\alpha}} + 100 \tag{66}$$

with $\alpha = 0.5$, n is the electrons density and n_{ref} is the reference electron density = 10^{17} cm^{-3}.

An interesting point in GaAs, as it can be seen from **Figures** 7 and **8**, that for GaAs, the saturation velocity of electrons can be reached even at relatively 5 and 10 V. This makes GaAs faster in terms of electron mobility and response time.

The fact that GaAs reaches saturation velocity at lower voltages means that the material can achieve high-speed operation without requiring high bias voltages. Lower bias voltages lead to reduced power dissipation, as power is related to voltage and current. With a smaller voltage required to achieve high-speed operation (due to the lower threshold for saturation velocity), the current also does not have to increase excessively, leading to a reduction in the overall power consumption. This is especially beneficial in high-frequency applications, where power efficiency is critical.

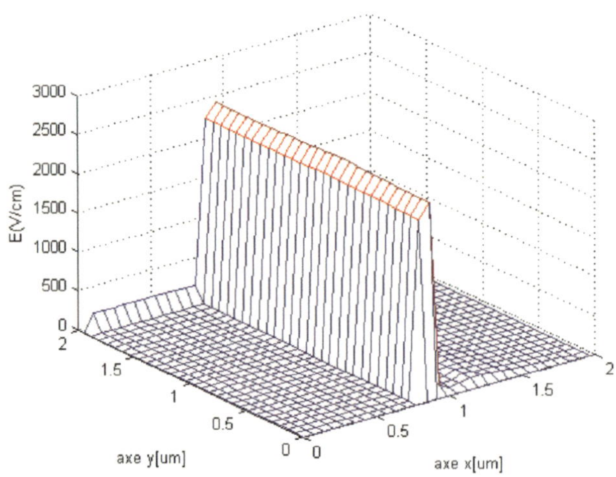

Figure 17.
The Electric Field profile at 5 volts.

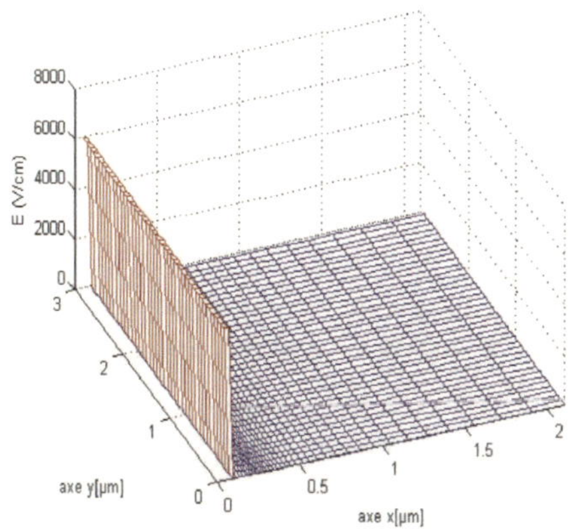

Figure 18.
The Electric Field profile at 10 volts.

The electric field increases considerably (**Figures 17** and **18**), and a different, trapezoidal form of the field is observed, which differs from that at weak injection and without illumination [39].

At final, using GaAs in a pin photodiode needs smaller bias voltages to respond efficienctly. That is a good point.

The reverse current of a PIN diode depends on the minority carrier densities in the neutral regions of the diode, as well as on the generation of electron-hole pairs in the space charge region, as previously explained. Radiation enhances the reverse current, as incident photons generate carriers in each region. The behavior of these free carriers varies depending on their location of creation, with diffusion or generation contributing to the photocurrent **Figures 19** and **20**.

The photocurrent in the PIN diode closely follows the conduction current at low injection levels, with an added generation current component. This is reflected in the curves, which are shifted upwards and do not originate from zero.

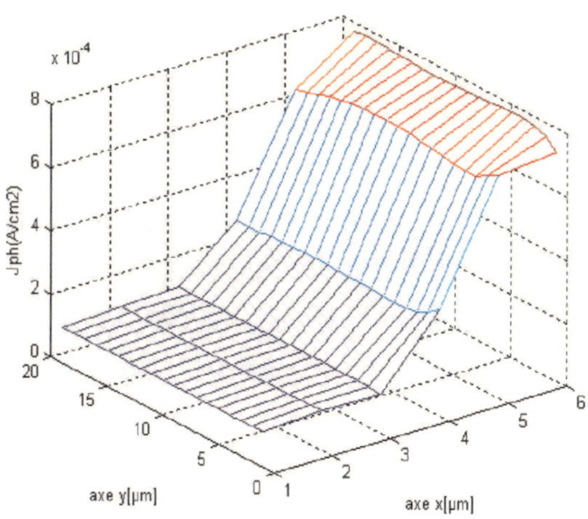

Figure 19.
The photocurrent profile at 5 volts.

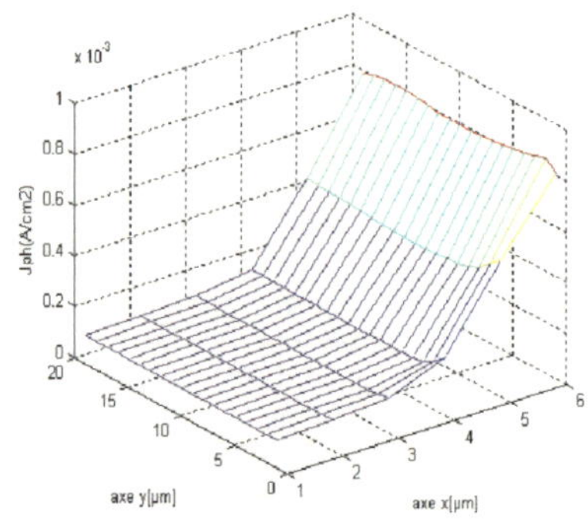

Figure 20.
The photocurrent profile at 10 volts.

12. Electrical characteristics of the photodiode

The PIN photodiode can be represented by a current source I_{ph}, which is due to the generation of electron-hole pairs that constitute the current. In parallel with the source, there is the junction capacitance and a high-value shunt resistor Rsh, which describes the leakage current. Additionally, there is an internal resistance of the photodiode, denoted Rs, which is in series. **Figure 21** shows the equivalent circuit of the photodiode.

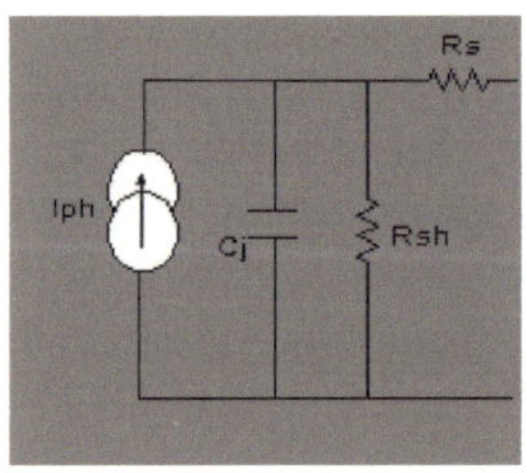

Figure 21.
The photodiode equivalent circuit.

- *Current Source (Iph):* This represents the photocurrent generated by the photodiode due to incident light.
- *Junction Capacitance (Cj):* This models the capacitance of the diode junction, which is typically a function of the reverse bias voltage.
- *Shunt Resistor (Rsh):* This represents leakage currents in parallel with the photodiode, usually having a very high value.
- *Series Resistor (Rs):* This accounts for the internal resistance of the photodiode, typically in series with the junction.

12.1 Temporal response of the photodiode

Three time constants are present in the temporal response of a photodiode, each corresponding to a distinct physical process:

- The lifetime constant associated with the diffusion current (10^{-8} s in GaAs), which corresponds to a bandwidth of 100 MHz. In many applications, such as telecommunications, this time constant is often too large to be acceptable, as it limits the speed of the photodiode response.

- The capacitive time constant, where the space charge region acts as a capacitance, denoted by C_d or C_j, and is described by the following relation:

$$C_j = A.\sqrt{\frac{q\varepsilon N_D}{2(V_{bi} - V)}} = \frac{A\varepsilon}{W} \quad (67)$$

- This time constant is influenced by the capacitance of the junction and the resistance of the circuit. It determines how quickly the photodiode can charge and discharge in response to changes in the incident light, affecting the high-frequency behavior and bandwidth of the device.

- The transit time constant, associated with the movement of charge carriers across the photodiode. It represents the time it takes for electrons and holes to cross the depletion region under the influence of the electric field. The transit time is affected by the width of the depletion region and the mobility of the carriers, and it plays a significant role in the high-speed performance of the photodiode, especially in fast switching applications.

For high-speed operation, the capacitance, in series with a series resistance of Rs = 50 Ω imposes a frequency limitation given by $1/2\pi RC$.

For a typical detector with an area of 1 mm^2, doped to $10^{15} cm^{-3}$ and biased at 10 V, the depletion region width is 3.6 μm, and the capacitance is 30 pF (as shown in **Figure 22**), charged through a 50 Ω resistance. This detector exhibits a bandwidth limitation, with a corresponding cutoff frequency of 110 MHz (as shown in **Figure 23**).

In general, this capacitance is approximately 100 pF at low bias voltages and decreases to a few tens of pF at higher bias voltage.

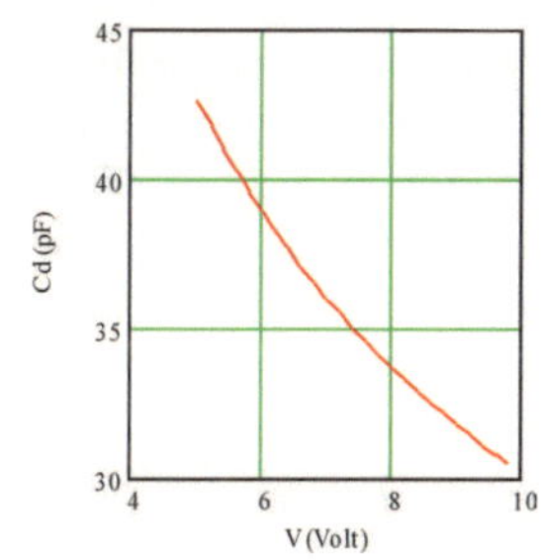

Figure 22.
The capacity variation versus the reverse bias.

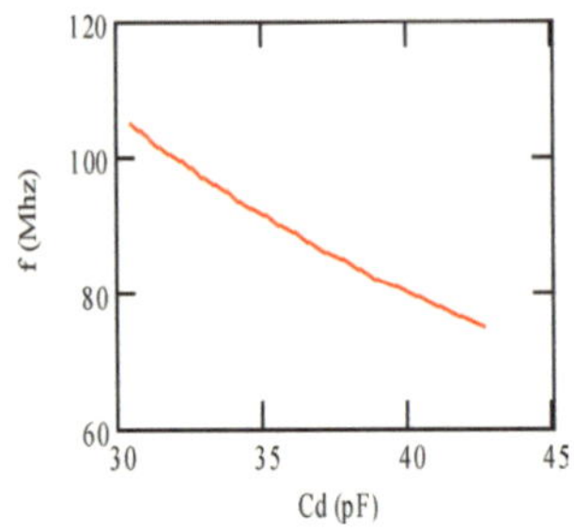

Figure 23.
The cutoff frequency as a function of capacitance.

13. Future work and recommendations

As a direction for future work, it is strongly recommended to explore the integration of perovskites due to their remarkable properties, which make them promising candidates for optoelectronic applications, such as PIN photodiodes. In addition, a comprehensive optical communication system involves not only optical fibers but also laser diodes and photodiodes. Therefore, a broader approach should focus on investigating various types of PIN photodiodes made from different materials and analyzing the entire communication path, from the emitter to the receiver.

In one of our calculations, transition metal doping was found to influence the electronic structure and magnetic properties of GaAs, which may enable enhanced control of electron spin for spintronic devices. This approach could potentially lead to the development of advanced materials with improved performance for spin-based electronics applications.

14. Conclusion

The physical model used in this work is based on the drift-diffusion approach within the framework of Maxwell-Boltzmann statistics. The variables of this system (ψ, n, p) were carefully chosen to satisfy both mathematical requirements (continuity, differentiability, etc.) and computational needs (handling of magnitude orders, precision in calculations, etc.).

The developed simulator allows the simulation of component conduction based on relaxation semiconductor devices. The program studies the transport phenomena in a PIN diode under illumination.

The program, written in C++, allows for the flexible and efficient usage of three-dimensional matrices. It employs a combined Gummel-Newton method for solving the nonlinear equations resulting from the physical model. Newton's method is well-suited for solving nonlinear equations, but it requires careful choice of initial values for convergence. To improve the control of computing and convergence, we have implemented a combined method that integrates both the Newton and Gummel methods. This creates a relay between these two algorithms, enabling better control of the simulation process.

The principle of the combined method is well known: Gummel's method is used initially to calculate numerical results in thermal equilibrium. The combined method has been applied to relaxation semiconductor devices under high trap density, illumination, and reverse bias conditions.

Introducing a high trap density can cause the system to diverge. To prevent this, we start the simulation with a moderate trap density. The results are then used as initial values for a new simulation with a higher trap density.

It is important to remember that for Newton's iteration to converge, the initial guess must be as close as possible to the true solution. As the iteration progresses, the residuals should decrease quadratically, indicating convergence. The Newton method typically converges quickly if the initial conditions are reasonably close to the solution and if the solution is unique in the region of interest.

Under reverse bias, the electron concentration values decrease significantly compared to those at thermal equilibrium (Gummel method). These values cannot be directly used as initial values in the Newton method. To overcome this issue, the free electron densities obtained from thermal equilibrium are scaled down by a factor of 10^8. This scaling ensures that the minority carriers are normalized, allowing them to be used as initial values for the Newton algorithm without the risk of extended computation time.

Thus, the combined method provides a balance between convergence speed, computational efficiency, and versatility, ensuring short run times and reliable results for relaxation semiconductor devices under high trap density and reverse bias conditions.

In conclusion, this approach improves our understanding and simulation of the behavior of PIN diodes under illumination and reverse bias while providing a robust solution for semiconductor devices in complex environments. The developed model represents a significant advancement in the simulation of semiconductor devices and could be extended to other types of optoelectronic components.

This chapter reviewed the theoretical foundations of conductivity models in semiconductor devices under illumination, with a focus on methods for handling time dispersion in capacitances in PIN devices. It also emphasized the importance of addressing traps and improving the accuracy of simulations in real-world environments. The results indicate that enhancing capacitance and the ability of devices to operate at higher speeds is crucial for future applications in optical fiber networks. Research into the effects of time dispersion and traps in PIN devices will open new horizons for developing optimized technologies to improve optical network performance.

Acknowledgements

I would like to express my sincere gratitude to my supervisor, the late Professor BRAHIMI MENEZLA Rabea, to whom I pay tribute for her support and valuable guidance. I also extend my thanks to the research team, particularly Dr. BENICHOU Boucif at the Materials Elaboration and Characterization Laboratory, University Djillali Liabès of Sidi Bel Abbès, DZ-22000, Sidi Bel Abbès, Algeria.

Conflict of interest

The authors declare no conflict of interest.

Notes/thanks/other declarations

I would like to thank the co-authors of this work for their contributions. I also deeply appreciate the unwavering support and encouragement from my family throughout the research process.

Author details

Badra Bouabdallah[1], Saliha Kheris[2*] and Zakia Nabi[1]

1 Computational Materials Physics Laboratory (CMPL), Department of Materials and Sustainable Development, Faculty of Exact Sciences, University Djillali Liabès of Sidi Bel Abbès, Sidi Bel Abbès, Algeria

2 Condensed Matter and Sustainable Development Laboratory (LMCDD), Department of Materials and Sustainable Development, Faculty of Exact Sciences, University Djillali Liabès of Sidi Bel Abbès, Sidi Bel Abbès, Algeria

*Address all correspondence to: karima_kh5@yahoo.fr

References

[1] Neudeck PG, Kleine JS, Sheppard ST, McDermott BT, Bedair SM, Cooper JA, et al. Electrical characterization of GaAs PiN junction diodes grown in trenches by atomic layer epitaxy. Applied Physics Letters. 1991;**58**(1):83-85

[2] Kobayashi KW, Oki AK, Tran L, Streit DC. A novel baseband-1.5 GHz monolithic HBT variable gain amplifier with PIN diode gain control. In: IEEE 1995 Microwave and Millimeter-Wave. Monolithic Circuits Symposium. Orlando, FL, USA: Digest of Papers; 1995. pp. 181-184. DOI: 10.1109/MCS.1995.470963

[3] Francoeur S, Seong MJ, Mascarenhas A. Applied Physics Letters. 2003;**82**(22):3874

[4] Longeaud C, Kleider JP, Kaminski P, Koslowski R, Pawlowski M, Cwirko J. Investigation of defect levels in semi-insulating materials by modulated and transient photocurrent: Comparison of methods. Semiconductor Science and Technology. 1999;**14**:747-756

[5] Longeaud C, Kleider JP, Kaminski P, Koslowski R, Pawlowski M, Cwirko R. New techniques for the characterization of defect levels in semi-insulating materials. In: IEEE SIMC Proceedings 10th Conference on Semi-Insulating Materials (SIMC-X), June 1–5, 1998, Berkeley, CA, USA. 1998. p. 76

[6] Sze SM, Ng KK. Physics of semiconductor devices. 3rd ed. Hoboken, NJ: Wiley-Interscience; 2006

[7] Shur MS. Physics of Semiconductor Devices. USA: Prentice Hall; 2012. ISBN 978-0470-53794-7

[8] Haug H, Koch S. Quantum Theory of the Optical and Electronic Properties of Semiconductors. 3rd ed. Singapore: World Scientific Publishing and Imperial College Press; 2004

[9] Liu HC, Chien CL. Bandgap and electron mobility in GaAs at 300 K. Journal of Applied Physics. 1992;**72**(7): 3675-3682

[10] Sze SM. Semiconductor Devices: Physics and Technology. United States of America: Wiley; 1981. ISBN 0-471-33372-7

[11] Schock HW, Gärtner E. The bandgap and electronic properties of germanium. Journal of Applied Physics. 1998;**65**(8): 3219-3222

[12] Lichtensteiger H, Bardeen J. Carrier mobility in GaAs and Si. IEEE Transactions on Electron Devices. 1990; **37**(12):2650-2656

[13] Bass M, Spector J. Velocity saturation effects in GaAs: Implications for high-speed devices. IEEE Transactions on Electron Devices. 2002;**49**(7):1180-1185

[14] Green MA. Third-generation photovoltaic solar cells. Solar Energy Materials and Solar Cells. 2001;**61**(1–2): 1-16

[15] Delaye P, Roosen G. Matériaux photoréfractifs. Collection SFO. 2003;**8**: 151-170

[16] Begotti M, Longo M, Magnanini R, Tarricone L, Gombia E, Mosca R, et al. Electrical and photoelectrical properties of a GaAs-based p-i-n structure grown by MOVPE. Crystal Research and Technology. 2005;**40**(10–11):1033-1038

[17] GeePS. Modulated photocurrent spectroscopy (MPC) [Internet].

Available from: https://www.geeps.centralesupelec.fr/index.php?page=pv_PCM [Accessed: February 28, 2025]

[18] Blaha P, Schwarz K, Madsen GKH, Kvasnicka D, Luitz J. WIEN2k, an Augmented Plane Wave plus Local Orbitals-Program for Calculating Cristal Properties. Vienna: Vienna University of Technology; 2001

[19] EMIS Datareviews Series N°08. Properties of Indium Gallium Arseride. London: INSPEC; 1991

[20] Castagne R, Duchemin JP, Gloanec M, Rumelhand C. Circuits intégrés en Arséniure de Gallium. Paris: Masson; 1989

[21] Cook RK. Numerical simulation of hot-carrier transport in silicon bipolar transistors. IEEE Transactions on Electron Devices. 1983;**ED-30**(9): 1103-1110

[22] Moreau Y, Manifacier JC, Henish HK. Minority carrier injection into relaxation semiconductor. Journal of Applied Physics. 1986;**60**:2904

[23] Rafferty CS, Pinto MR, Dutton W. Iterative methods in semiconductor device simulation. IEEE Transactions on Electron Devices. 1985;**ED-32**(10): 2018-2027

[24] Johnston SW, Ahrenkiel RK, Friedman DJ, Kurtz SR. Deep-level transient spectroscopy in InGaAsN lattice-matched to GaAs. Conference Record of the Twenty-Ninth IEEE Photovoltaic Specialists Conference, 2002., New Orleans, LA, USA, 2002. pp. 1023-1026. DOI: 10.1109/PVSC.2002.1190779

[25] Shockley W, Read WT. Statistics of the recombinations of holes and electrons. Physical Review. 1952;**87**(5): 835-842

[26] Mathieu H. Physique des semiconducteurs et des composants électroniques. Quatrième ed. Paris: Masson; 1998

[27] Saidane A. Physique des semiconducteurs. Algerie: Office des Publications Universitaires; 1993

[28] Mathieu H, Fanet H. Physique des semi-conducteurs et des composants électroniques, cours et exercices corrigés. Dunod. Paris. 2009. ISBN: 978-2-10-051643-8

[29] Lévy F. Physique et technologie des semi-conducteurs. Presses polytechniques et universitaires romandes. Lausanne, CH-1015. 1995. ISBN: 2-88074-272-2

[30] Dimitrijev S. Principles of Semiconductor Devices. 2nd ed. NewYork: Griffith University, Oxford University Press; 2012. ISBN 978-0-19-538803-9

[31] Sapoval B, Hermann C. Physics of semiconductors. Berlin: Springer-Verlag; 1995

[32] Neamen DA. Semiconductor Physics and Devices. 3rd ed. New York: Published by McGraw-Hill; 2003

[33] Li SS. Semiconductor Physical Electronics. 2nd ed. Springer; 2006. ISBN 10: 0-387-28893-7

[34] Ardebili R. Étude par simulation numérique des phénoménes de transport dans les semiconducteurs à relaxation [Thése de doctorat, Université de Montpellier II]. France. 1992

[35] Boisrobert C, Toffano Z. Cours de DESS et DEA d'Optique et

Optoélectronique. Nantes, France: Université de Nantes; 2001

[36] Berthier P. Transistors à effets de champs AlInAs/(Al)GaInAs(P) pour photodétection intégrée à 1,3–1,5 μm [thèse de doctorat]. Montpellier, France: de l'Université de Montpellier II; 1993

[37] Aupetit-Berthelemot C. Etude des effets parasites du transistor à haute mobilité électronique (HEMT) sur InP pour applications micro-optoélectroniques [thèse de doctorat]. Limoges, France: de l'Université de Limoges; 1998

[38] Cassan É. Une introduction aux télécommunications optiques par la simulation de systèmes simples. Journal sur l'enseignement des sciences et technologies de l'information et des systèmes (J3eA). 2003;**2**. Article 007. DOI: 10.1051/j3ea:2003007

[39] Bouabdallah B, Bourezig Y, Brahimi R. The investigation of a novel approach in the simulation of the trap levels effect on the conduction in GaAs P +vN+ diode. Journal of Materials Processing Technology. 2009;**209**: 1495-1503

[40] Franz AF, Franz GA, Selberherr S, Ringhofer C, Markowich P. Finite-boxes: A generalization of the finite-difference method suitable for semiconductor device simulation. IEEE Transaction on Electron Devices. 1983;**ED-30**(9): 1070-1082

[41] Gummel HK. A self-consistent iterative scheme for one-dimensional steady-state transistor calculations. IEEE Transaction on Electron Devices. 1964; **ED-11**:455-465

[42] Gokhale BV. Numerical solutions for one-dimensional silicon n-p-n steady-state transistor calculations. IEEE Transactions on Electron Devices. Aug. 1970;**ED-17**(8):594-601. DOI: 10.1109/T-ED.1970.17043

[43] Demari A. An accurate numerical one-dimensional solution of the p-n junction under arbitrary transient conditions. IEEE Solid-State Electronics Pergamon Press. 1968;**11**:1021-1053

[44] Hachtel GD, Joy RC, Cooley JW. A new efficient one-dimensional analysis program for junction device modeling. Proceedings of the IEEE. Jan. 1972;**60**(1) 86-98. DOI: 10.1109/PROC.1972.8561

[45] Dhatt G, Touzout G. Une présentation de la méthode des éléments finis. Paris: Collection Université de Compiègne-Malonie S.A. éditeur; 1984

[46] Demari A. An accurate numerical steady-state one-dimensional solution of the pn junction. Solid-State Electronics. 1968;**11**:33-53

[47] Buturla EM, Cottrell PE. Simulation of semiconductor transport using coupled and decoupled solution techniques. Solid-State Electronics. Pergamon Press. 1980;**23**:331-334

[48] Kemp M, Tannous CG, Meunier M. Amorphous silicon device simulation by an adapted Gummel method. IEEE Transaction on Electron Devices. 1988; **35**(9):1510-1513

[49] Akcasu OE. Convergence properties of Newton's method for the solution of the semiconductor transport equations and hybrid solution techniques for multidimensional simulation of VLSI devices. Solid-State Electronics. 1987; **27**(4):319-328

[50] Buturla EM, Cottrell PE, Grossman BM, Salsburg KA. Finite-element analysis of semiconductor devices: The FIELDAY program.

IBM Journal of Research and Development. juil. 1981;**25**(4):218-231. DOI: 10.1147/rd.254.0218

[51] Selberherr S. Analysis and Simulation of Semiconductor Devices. Wien, New York: Springer-Verlag; 1984

[52] Wu KC, Lucas RF, Wang ZY, Dutton RW. New approaches in a 3-D one carrier device solver. IEEE Transient Computer-Aided Design of ICAS. 1989; **CAD-8**(5):528-537

[53] Viallet JE, Mottet S. Heterojunction under Fermi-Dirac statistics: General state equations and steady state numerical methods. In: Miller JJH, editor. Proceedings of NASECODE IV Conf. Dublin: Book Press; 1985. pp. 530-535

[54] Menezla R. Réalisation d'un logiciel de résolution de l'équation de poisson à trois dimensions : Simulation numérique tridimensionnelle du claquage des composants à jonctions P-N, Thése de Doctorat d'Etat. France: Ecole Centrale de Lyon; 1990. pp. 85-05

[55] Ruch JG, Kino GS. Transport properties of GaAs. Physics Review. 1968;**174**:921-931

[56] Walukiewicz W, Lagowski L, Jastrzebski L, Licteusteiger M, Gatos H. Electron mobility and free carrier absorption in GaAs: Determination of the compensation ratio. Journal of Applied Physics. 1979;**50**:899-908

Chapter 5

Optimal Photodeposition of Co-Catalyst Over 2D Heterostructures for Improved H_2 Production/Water-Splitting

Durre Sameen, Saqib Ahmad and Faryal Idrees

Abstract

The resources of fuel have been reduced to a dangerous level. In the upcoming few years, these resources will come to an end. This threat has created an alarming situation for upcoming generations. Therefore, we need alternative energy resources for energy production, such as evolution of hydrogen from water, solar energy, and from biomass. This chapter provides a comprehensive review of advanced methodologies focused on enhancing hydrogen production from water splitting. This chapter reveals the modification of bandgap, enhanced stability, and extensive surface area of co-catalyst photodeposition on 2D heterostructures for photocatalytic performance. This chapter primarily focuses on strategic heterostructure doping with transition metals, including silver, gold, platinum, *etc.*, as well as the optimization of reduction potential. These modifications enhance charge separation and reduce recombination rates for modified photocatalytic efficiency. Results demonstrate that precise loading of co-catalysts and optimized doping conditions enhance the production of hydrogen. The chapter highlights an extensive overview of synthesis, characterization techniques, and integrating theoretical insights with experimental data. These findings contribute to finding the solution of sustainable energy solutions through the transformative potential of co-catalysts and doped 2D heterostructures to achieve higher efficiency of hydrogen production.

Keywords: photocatalysis, water splitting, H_2 production, heterostructures, photodeposition

1. Introduction

2D heterostructures exhibit fascinating electronic and optical characteristics, which make them highly promising for various applications in catalysis and energy conversion. Their ability to tune bandgaps and efficiently separate charges greatly enhances catalytic performance, particularly in photocatalytic water splitting for

IntechOpen

hydrogen production. Photocatalytic water splitting (PCWS) using solar energy has gained attention as a viable and sustainable method to address energy challenges. The progress of this technology largely hinges on the development of high-performance photocatalytic materials. Therefore, identifying and designing efficient photocatalysts has become a central objective in the fields of photocatalysis and renewable energy [1]. In addition to size, dimensionality significantly influences material properties. For example, carbon-based materials like carbon nanotubes, graphite, graphene, and fullerenes display distinct physical and chemical traits as 0D, 1D, 2D, and 3D structures, respectively. The ground-breaking discovery of graphene by Novoselov et al. [2] made it possible to study and design various 2D materials. 2D catalyst materials have the ability to increase stability, selectivity, and mobility in catalysis. According to both theoretical and experimental research, ultrathin 2D materials have significant catalytic benefits because of their enormous exposed surface area, which minimizes thickness and enhances interactions with reactants. Furthermore, exposing inner atoms on the surface through thinning material to a monolayer increases the number of catalytic active sites. In addition to introducing defects, unsaturated sites, and active edges, this thickness decrease may also improve ion mobility and mass transfer, reduce energy barriers for adsorption-desorption, and increase catalytic efficiency [3]. Due to its unique attributes, which are essential for many applications, such as photocatalysis for the creation of hydrogen, two-dimensional (2D) heterostructures constitute a substantial breakthrough in material research. Their atomic thickness and large surface area, which prevent the formation of non-selective sites and increase catalytic activity and selectivity, make them unique.

2. Significance and properties of 2D heterostructures

2D heterostructures are innovative materials prepared by layering or combining different two-dimensional (2D) substances, such as graphene, transition metal dichalcogenides (TMDs) like MoS_2 and WS_2, hexagonal boron nitride (hBN), or black phosphorus. These structures provide an exceptional opportunity to study and customize new electronic and optical properties, positioning them as key building blocks for advancing modern nanotechnology and next-generation applications. 2D heterostructures are revolutionary materials formed by stacking different two-dimensional layers, such as graphene, transition metal dichalcogenides (TMDs), or hexagonal boron nitride (hBN). These materials offer unmatched flexibility because their layers are held together by weak van der Waals forces, allowing precise stacking without the need for lattice matching. By combining properties like the high carrier mobility of graphene with the bandgap features of TMDs, 2D heterostructures overcome the limitations of traditional bulk materials, making them ideal for high-performance electronics and optoelectronics. Their unique characteristics, such as tunable band structures, quantum effects, and interlayer coupling, enable the customization of energy gaps and other properties, making them highly versatile for applications ranging from quantum computing and flexible electronics to photovoltaics and advanced sensors. Many highly efficient semiconductors, such as metal oxides, have been investigated to produce hydrogen through water splitting, but their efficiency is still far below the line of practical application. In recent times, people have taken a keen interest in 2D heterostructures with a suitable band gap which consists of several semiconductors that overcome the drawbacks of a single semiconductor. 2D materials inherently possess a high surface-to-volume ratio, which produces more active sites and improves photocatalytic reactions by increasing interaction with water molecules. The electronic properties of 2D materials can easily be tuned. The addition of a co-catalyst plays a very important role in the photocatalytic reaction of heterostructure by facilitating the separation of photogenerated electron-hole pair and increasing the overall efficiency of water splitting. Co-catalysts promote redox reactions at the surface [4]. The preparation of co-catalysts from noble metals like Au and Pt to non-noble metals has been a great site of interest for researchers but their stability, cost, and scarcity prohibit us from using them [5]. Transition metal-based co-catalysts like cobalt and nickel have also been discussed but their efficiency significantly reduces with time. Metal oxide-based co-catalysts do not transfer the charge carrier efficiently. Therefore, we need the type of materials that are cost-effective, highly stable, possess high quantum efficiency, and are present in abundant amounts on Earth. 2D materials present a new class of emerging materials in energy storage and conversion applications, with unique thickness-dependent properties in the optoelectronic field, distinct charge mobilities, and larger surface areas [6]. This two-dimensional heterostructure indicates one of the great leaps forward in materials science, affording unique properties crucial in many applications, including photocatalysis for hydrogen production. 2D heterostructures have recently been considered highly important in material science due to their unique properties and huge potential in various fields like catalysis, electronics, and energy conversion. More precisely, due to the fascination with their capability to promote hydrogen evolution using water-splitting reactions, positioning these heterostructures as key materials for solving major energetic challenges around the world is in order. These materials provide flexible alternatives for sustainable H_2 generation (**Figure 1**) [7].

2.1 Physical and chemical properties of 2D materials

2.1.1 High surface area

2D materials have a very high surface-to-volume ratio due to which more active sites are exposed for catalytic reactions. Due to their exceptional physicochemical characteristics, two-dimensional (2D) heterostructures are leading the way in research for hydrogen production by water splitting. High surface areas, which are essential for promoting the photocatalytic processes required for effective hydrogen

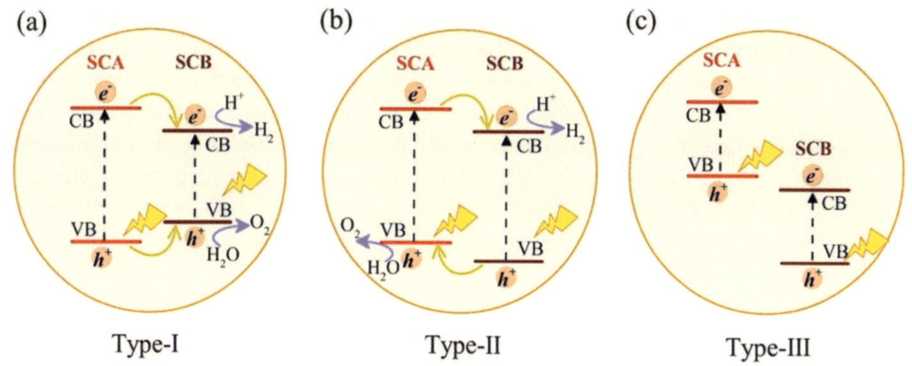

Figure 1.
Schematic representation of the separation of electron-hole pairs in traditional light-responsive heterostructures (HSs): (a) type I, (b) type II, and (c) type III, reproduced with permission [1].

production, are generally advantageous for these structures. Many active sites are provided by the large surface area of 2D heterostructures, which is necessary for the adsorption of water molecules and the subsequent production of hydrogen [8]. Enhancing charge separation and lowering electron-hole recombination are the main benefits of 2D heterostructures in hydrogen production. This is crucial because efficient charge separation makes the photocatalytic water-splitting process more efficient. For these purposes, materials such as transition metal dichalcogenides (TMDs) and graphitic carbon nitride (g-C3N4) have been extensively researched. In addition to having large surface areas, these materials have appropriate band edge locations that promote hydrogen evolution and water molecule activation [7, 9]. The Langmuir-Hinshelwood kinetics can be used to mathematically express the link between surface area and photocatalytic activity where the amount of reactant coverage on the catalyst surface determines the reaction rate:

$$r = k\theta_A\theta_B \tag{1}$$

Here, r and k represent the reaction rate and rate constant, θ_A and θ_B are the surface coverage of reactants A and B, respectively [9].

The efficiency of materials such as MoS_2, WS_2, and layered double hydroxides in photocatalytic hydrogen production has been particularly highlighted due to their ability to optimize the physicochemical properties through surface functionalization and bandgap engineering. These changes greatly improve their photocatalytic activity by enhancing charge carrier dynamics and light absorption, which are essential for effective solar-to-hydrogen conversion [8].

2.1.2 Bandgap tunability

In 2D heterostructures, band gap tunability refers to adjusting the difference of energy between the valence and conduction band. This can be done by changing the doping concentration, applying strain, making heterostructure with other suitable 2D material, or varying the number of layers of 2D material that leads to the quantum confinement effect. For HER, the input energy required is 0 eV with respect to the normal hydrogen electrode while for OER, this energy is 1.23 eV. The conduction band minimum of the 2D heterostructure should be more negative than the H^+/H_2 (i.e. 0.0 eV) redox potential and the valence band should be more positive than O_2/H_2O (i.e. 1.23 eV). The band gap directly affects the optical properties by allowing the material to absorb more visible light due to its narrowing. Narrowing of band gap also improves the conductivity of charge carriers but it can also increase the recombination of charge carriers. A wider band gap causes the effective separation of photogenerated charge carriers which results in effective photocatalytic activity. The electrical characteristics of 2D materials can be modified by altering their structures and compositions or through external stimuli. Due to its tunability, band alignments for certain processes can be optimized. Strain can affect the bandgap of 2D materials Eg as [6]:

$$E_g(\epsilon) = E_g(0) - \beta \times \epsilon \tag{2}$$

Where $E_g(0)$ represents the bandgap at zero strain and β represents the material-specific constant.

Because of their large surface area and adjustable bandgap, 2D heterostructures have emerged as a key component in effective hydrogen production by water splitting. The bandgap of these materials is important as it determines how efficiently the light is absorbed and how efficiently the charge carriers are generated and separated [10].

2.1.2.1 Bandgap engineering

In 2D heterostructures, bandgap engineering enables the optimization of light absorption throughout the solar spectrum. Because of their tunable electrical characteristics, which can be precisely adjusted by doping, applying strain, and creating heterojunctions, materials like MoS_2, WS_2, and g-C_3N_4 are widely used. For example, shifting the bandgap through the application of external strain or the introduction of transition metals can improve photocatalytic activity and solar light absorption [11]. Eq. (3) represents the connection between bandgap energy (Eg) and particle size or strain of quantum dots.

$$Eg = Eg0 + \frac{h^2}{8\pi^2}\frac{1}{m^* r^2} \tag{3}$$

Here, E_{g0} represents the bulk bandgap energy, h is Planck's constant, m^5 is the effective mass of the electron, and r is the radius of the quantum dot [11].

2.1.2.2 Material examples and their significance

Due to their stability and rich surface chemistry, MoS_2 and WS_2 are important for photocatalytic water splitting. High catalytic activity and efficient charge transport are made possible by their multilayer structure [12, 13]. Because of its polymer-like structure, graphitic carbon nitride (g-C_3N_4) offers a strong foundation for photocatalytic applications, especially when exposed to visible light [14]. There are more active sites available for photocatalytic reactions since 2D materials have a large surface area by nature. This characteristic is essential for increasing the pace at which hydrogen is produced because it permits a greater degree of interaction between the catalyst and water molecules, which raises the photocatalytic efficiency [15]. Although using 2D heterostructures to produce hydrogen has advanced significantly, there are still obstacles to overcome to achieve high quantum efficiency and scalability. Future studies are focused on creating novel heterostructures with ideal band alignment, creating new 2D materials with exceptional stability, and creating scalable and cost-effective methods [1]. In order to fulfill global energy demands, these discoveries and techniques are opening the door for the creation of next-generation photocatalytic devices that can use solar energy to produce hydrogen sustainably [6].

2.1.3 Enhanced charge carrier mobility

High electron and hole mobilities are common in 2D materials, which lower recombination energy losses in photocatalytic reactions. The mobility μ is calculated by Eq. (4) [16].

$$\mu = e\tau m^* \tag{4}$$

Where e is the charge on the electron, τ represents the scattering time, and m^* is the effective mass of charge carriers. In 2D materials such as graphitic carbon nitride ($g\text{-}C_3N_4$) and transition metal dichalcogenides (TMDs), charge carrier mobility is an essential component. Effective light absorption and conversion into chemical energy depend on the quick passage of charge carriers across the material, which is made possible by it. The overall rate of hydrogen creation is increased by high mobility, which ensures that electrons produced during light absorption immediately reach the surface to take part in the hydrogen evolution reaction. In order to efficiently fabricate 2D heterostructures, recent research has investigated a variety of coating and printing methods. To sustain constant photocatalytic activity over wide areas, uniform and continuous films are made using methods like spray casting, inkjet printing, and 3D printing [7]. The incorporation of 2D heterostructures into scalable hydrogen production systems is being made possible by the ongoing advancements in their synthesis and application methods. Considerable improvements in the efficiency of hydrogen production are expected as a result of continued research aimed at improving the charge carrier dynamics and structural characteristics of these materials [16].

2.2 Significance in water splitting

The development of hydrogen, especially through the water-splitting process using two-dimensional (2D) heterostructures, has gained attention as a result of the search for sustainable energy options. These materials offer significant advantages due to their atomically thin properties and high surface area for water splitting by photocatalysis. The number of active sites available for enhancing the hydrogen evolution reaction (HER) is directly correlated with the high surface area of 2D heterostructures. This property is essential because it permits a wider range of interaction between the catalyst and water molecules, which promotes higher rates of water dissociation and the subsequent generation of hydrogen [17]. The overall water-splitting reaction is expressed by Eq. (5).

$$H_2O \rightarrow H_2 + 1/2\,O_2 \quad \Delta G(298k) = 237.1\,kj\,mol^{-1} \tag{5}$$

The input energy requires a thermodynamic voltage of 1.23 V. Without bias, the conduction band minimum of the semiconductor photocatalyst should be more negative than the H^+/H_2 redox potential and the valence band should be more positive than O_2/H_2O. When a photon with energy equal to hυ or greater than the band gap energy of the semiconductor is absorbed, the electron in the valence band gets excited and moves into the conduction band and positive holes are left behind in the valence band. Generated electron takes a part in H_2 evolution at the cathode and the hole helps in O_2 evolution at the anode followed by the following reaction.

Hydrogen evolution reaction is represented in Eq. (6).

$$2H^+ + 2e^- \rightarrow H_2 \tag{6}$$

Oxygen evolution reaction is represented in Eq. (7).

$$H_2O \rightarrow ½O_2 + 2H^+ + 2e^- \tag{7}$$

In 1972, for the very first time, Honda and Fujishima first investigated water splitting through photoelectrochemical reaction using TiO_2 crystal as a photoanode and Pt as a cathode. The reaction rate for hydrogen production can be expressed by Eq. (8).

$$\text{Rate} = \text{k}\cdot\exp\left(-\frac{E_a}{RT}\right) \tag{8}$$

Where k is the reaction rate constant, E_a is the activation energy, R is the universal gas constant, and T is the temperature. The availability of more active sites effectively reduces E_a, thereby increasing the reaction rate [18].

2.2.1 Materials and their photocatalytic roles

Graphene is known for its exceptional electrical conductivity and mechanical strength, graphene serves as an excellent support material in heterostructures, improving the electron mobility and durability of the catalyst [12]. Graphene and molybdenum disulfide (MoS2) are commonly combined to create heterostructures that improve photocatalytic activity by reducing recombination and improving charge separation. MoS_2 edge sites are active for hydrogen adsorption and desorption [14]. The polymer-like substance known as graphitic carbon nitride (g-C_3N_4) has found widespread application due to its stability under photocatalytic conditions and visible light sensitivity. g-C_3N_4 increases the range of light absorption when included in heterostructures, which speeds up the synthesis of hydrogen [12]. The effectiveness of these heterostructures is further increased by adjusting the interlayer spacing, strain engineering, and doping which tuned their electrical properties. Researchers can maximize hydrogen generation by adjusting these characteristics by optimizing the band alignment for more effective charge transfer and separation. Combining several 2D materials frequently produces synergistic effects, where the qualities of the composite are better than their individual properties. This effect is very useful because the complexity of the reaction mechanisms in water splitting requires an extremely effective catalytic system [3]. When considering photocatalytic water splitting, the characteristics of 2D heterostructures are very important for a number of reasons:

2.2.2 Enhanced light absorption

The remarkable light-absorbing properties of 2D heterostructures, which are ascribed to their distinct electronic structures, make them useful for solar-driven hydrogen synthesis. These materials have highly adjustable bandgaps that enable them to absorb maximum light by modifying the electrical characteristics through doping, compositional changes, or structural alteration [19]. Materials like WS_2, MoS_2, and graphene are well known to absorb a wide range of light, including visible and near-infrared (NIR) regions which is essential for effectively promoting the photocatalytic water-splitting process [20, 21]. Graphitic Carbon Nitride (g-C_3N_4), an extensively researched 2D material, shows excellent properties due to high stability and suitable band gap for water splitting when exposed to sunlight [22, 23].

2.2.3 Facilitated electron-hole separation

In heterostructures, the strategic band edge alignment promotes effective charge separation which is essential for enhancing the photocatalytic activity. The alignment of the valence band maximum (VBM) and conduction band minimum (CBM) across the heterojunction is often used to illustrate this. Effective electron-hole pair separation is essential to avoiding recombination, which is a common problem that drastically reduces photocatalytic system effectiveness. This separation is made easier by 2D heterostructures, especially those made of layered materials, due to their integrated electric fields and heterojunction interface. By facilitating rapid spatial separation of charge carriers, layered TMDs can create van der Waals heterostructures that limit recombination and enhance effective charge transport when combined with materials like graphene [17].

2.2.4 Strong mechanical stability

For 2D heterostructures to last and maintain their functional integrity in photocatalytic applications, their mechanical robustness is essential. Cross-layer linkage and integration with substrates that offer extra mechanical support improve the strength and flexibility of these materials. When combined with other 2D materials, flexible materials like graphene improve overall structural robustness without losing electrical or photocatalytic capabilities [17]. Commercial water-splitting devices by integration of 2D heterostructure are still required to overcome challenges related to scalability and cost-effectiveness. For the large-scale production of high-quality materials, synthesis process innovations like chemical vapor deposition (CVD) and exfoliation are essential [3]. 2D heterostructures provide a pathway to effectively capture solar energy and have great potential for sustainable hydrogen production. Their development is a major step toward the realization of practical and extensible hydrogen fuel production systems, which are essential for meeting the world's energy demands in a sustainable manner. To fully utilize these materials' potential and make sure they satisfy the performance and cost requirements required for broad use in the energy sectors, more research and development is required [3].

3. Basic principle of hydrogen production

This process involves the use of sunlight to split water into hydrogen and oxygen.

$$2H_2O + \text{sunlight} \rightarrow 2H_2 + O_2 \quad (9)$$

When sunlight is exposed to a 2D heterostructure, it generates electron and hole pairs. For efficient hydrogen production, it is necessary that these photogenerated charge carriers separate efficiently. Co-catalyst helps to separate these charge carriers. The generated electrons move to the conduction band of the photocatalyst and holes are left behind in the valence band which oxidizes water molecules and produces oxygen gas and hydrogen ions. The electrons have enough energy to further react with hydrogen ions and produce hydrogen gas followed by

$$2H^+ + 2e^- \rightarrow H_2 \quad (10)$$

4. Proposed methods to improve H_2 production

In the transition to sustainable energy systems, hydrogen production is essential, but conventional techniques often require a lot of energy or non-renewable resources.

Significant efforts are being made to increase the efficiency and environmental impact of hydrogen generation in response to the growing need for cleaner energy options. The development of photocatalytic materials, particularly the strategic application of co-catalysts over 2D heterostructures, has energized the search for effective hydrogen production by water splitting. In order to improve hydrogen production, this chapter explores the optimal photodeposition of co-catalysts on 2D materials, emphasizing important techniques and their key ideas [24]. Here is a thorough breakdown of the several approaches currently being used to increase H_2 generation:

4.1 Steam methane reforming (SMR)

For industrial hydrogen production, Steam methane reforming (SMR) is the most widely used and economically feasible technique. Methane and steam undergo a catalytic reaction at high temperatures to produce carbon monoxide and hydrogen. This approach is quite effective in terms of hydrogen generation and makes use of plentiful natural gas resources. However, due to the significant emission of carbon dioxide during this process, it also raises environmental challenges [25]. Steam and methane are combined, and the mixture is heated to temperatures ranging from 700 to 1000°C, while maintaining a pressure of 3–25 bar. This procedure involves an endothermic reaction to generate carbon monoxide and hydrogen:

$$CH_4(g) + H_2O(g) \rightarrow CO(g) + 3H_2(g) \quad (11)$$

Since nickel-based catalysts can withstand high temperatures and effectively catalyze the reforming reaction, they are frequently utilized to speed up the process. This method is the most prevalent for industrial hydrogen production, providing a high yield of hydrogen [26].

4.2 Electrolysis

Electrolysis is a transformative technology that allows for the decomposition of water into its elemental components, hydrogen and oxygen, using electrical energy. This method is not only fundamental but also pivotal in the shift toward sustainable and clean energy production. It decomposes water into oxygen and hydrogen using electricity [27]. An electric current is passed through water containing an electrolyte, causing hydrogen and oxygen gases to be released at the cathode and anode, respectively:

$$2H_2O(l) \rightarrow 2H_2(g) + O_2(g) \quad (12)$$

Noble metal catalysts like platinum and iridium are often used to enhance efficiency and to provide a clean hydrogen production route, especially when powered by renewable energy sources [28, 29].

4.3 Photocatalytic water splitting

Photocatalytic water splitting is a cutting-edge technique that leverages the power of sunlight to convert water into hydrogen and oxygen, presenting a promising pathway toward sustainable hydrogen production. This method utilizes solar energy to catalyze the splitting of water molecules. The energy absorbed from sunlight excites the electrons in the photocatalyst, creating electron-hole pairs that are essential for driving the chemical reactions needed to split water into its elemental components. Photocatalysts absorb photons, generating electron-hole pairs that drive the water-splitting reactions [30]. The efficiency (η) of photocatalytic water splitting can be described by:

$$\eta = \frac{\text{Power out}\,(\text{Hydrogen})}{\text{Power in}\,(\text{Solar})} \qquad (13)$$

The efficiency of the photocatalyst and the total solar-to-hydrogen conversion efficiency can be evaluated with the use of this formula. Common materials include TiO_2, CdS, and newly developed perovskites. This method harnesses solar energy, making it highly sustainable [27].

4.4 Biological production

Biological hydrogen production utilizes the intrinsic capabilities of certain microorganisms, such as algae and bacteria, to generate hydrogen, representing a renewable approach to energy generation. This method taps into the metabolic processes of these organisms, where biologically mediated reactions lead to hydrogen gas as a byproduct. Certain microorganisms have metabolic pathways that, through direct bio photolysis, indirect bio photolysis, or fermentation, produce hydrogen. These processes are catalyzed by specific enzymes such as hydrogenases and nitrogenases [31, 32]. These microorganisms are cultured under controlled conditions that optimize their growth and hydrogen production efficiency. Parameters such as pH, temperature, nutrient supply, and light intensity are meticulously managed to maximize yield. Advances in genetic engineering and biotechnology have led to the development of modified strains of bacteria and algae that demonstrate enhanced hydrogen production capabilities compared to their natural counterparts [32].

4.5 Thermochemical water splitting

Thermochemical water splitting represents a high-temperature chemical process designed to decompose water into hydrogen and oxygen, offering an alternative route to traditional electrolytic methods. This method capitalizes on the thermodynamic shifts that occur at elevated temperatures, enabling the efficient breakdown of water molecules through a series of chemical reactions. This method utilizes multiple thermochemical cycles, where heat energy is converted into chemical energy, facilitating the disassociation of water. The process typically involves two or more chemical reactions. Heat is supplied, often from solar, nuclear, or residual industrial sources, to drive endothermic reactions. These reactions are cyclically reversed to produce hydrogen and oxygen separately. High-temperature ceramics and other refractory materials are used to contain and manage the extreme conditions necessary for the reactions. Various chemical compounds, such as metal oxides, are also employed to mediate the reactions, undergoing reduction and oxidation in a cyclic manner [33]. This process offers a large-scale, continuous production capability and is especially attractive when combined with nuclear or renewable energy sources. Compared to fossil fuel-based techniques, thermochemical splitting can drastically lessen its environmental impact by utilizing high temperatures and potentially emission-free energy sources (**Figure 2**) [32, 33].

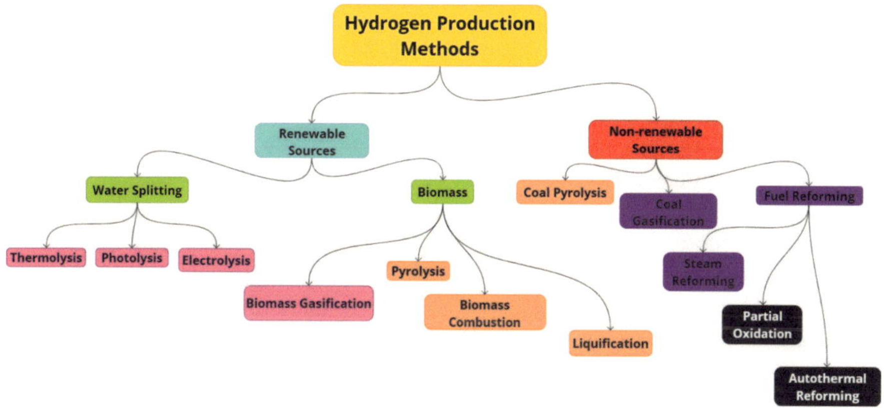

Figure 2.
Hydrogen production methods.

5. Processes involved in photocatalytic H_2 production

Photocatalytic hydrogen production utilizing 2D heterostructures involves a number of steps, all of which are essential to increasing the system's effectiveness [7].

5.1 Photocatalytic mechanisms

For efficient photocatalytic processes, 2D heterostructures are designed to maximize light absorption and improve charge separation. Band gaps in 2D materials, such as MoS_2 and WS_2, are ideal for absorbing significant portions of the solar spectrum. For example, MoS2 can efficiently absorb visible light due to its band gap, which varies between 1.2 and 1.9 eV depending on the thickness of its layer [34]. The optical properties of these materials are impacted by their thickness, leading to quantum confinement effects, which occur when electron mobility is restricted to two dimensions. As the material's dimensions reduce, this confinement may cause the band gap to expand, improving the material's ability to absorb light. Many 2D heterostructures, such as MoS_2/WS_2, form type-II heterojunctions when combined. The valence band maximum and conduction band minimum of the two materials in these heterojunctions overlap one another, resulting in an inherent electric field at the interface [1]. By separating electron-hole pairs produced by light absorption, this field lowers recombination rates and increases the lifetimes of charge carriers. Effective spatial separation of charges is facilitated by the heterojunction's energy band alignment. Differences in electrochemical potential cause electrons and holes to be driven in opposing directions, increasing separation efficiency, and decreasing recombination. The distinct layered structures of materials like graphene, MoS_2, and WS_2 facilitate effective charge dynamics, which are essential for photocatalytic activity. These materials have a high surface area to volume ratio due to their atomically thin and layered structure, which increases the number of active sites available for reactions and interactions with matter and light. For photocatalytic applications to effectively disperse energy and adsorb additional reactant molecules, this large surface area is essential. For instance, graphene's remarkable electrical conductivity is well-known. By acting as an electron sink or transport layer, graphene can help carry charge carriers from the photocatalyst surface to a reaction site when combined with other 2D materials such as MoS_2 or WS_2. This boosts the photocatalytic process's efficiency and lowers the possibility of electron-hole recombination [35].

5.2 Band alignment and charge separation

The precise alignment of the materials' valence band (VB) and conduction band (CB) often hinders the efficiency of photocatalytic devices. For instance, charge carriers can be effectively separated using type-II heterojunctions. A driving force that separates the electron-hole pairs and prolongs their lifespan is created in these systems when the CB and VB of one material are energetically lower than those of another, increasing the efficiency of hydrogen production [10]. The conduction band minimum (CBM) and valence band maximum (VBM) of one semiconductor material are positioned energetically lower than those of another semiconductor in contact with it in a Type-II heterojunction. Because this kind of band alignment separates charge carriers spatially, it is essential. There is an inherent electric field or potential gradient created by the offset between the CBM and VBM across the connection. This field pulls the photogenerated electrons and holes in opposite directions, acting as a driving force. Through spatial separation, the heterojunction considerably lowers the likelihood of electron and hole recombination. A frequent problem in photocatalytic systems is recombination, which lowers the effectiveness of charge carriers in promoting the intended processes, including the creation of hydrogen. Charge carriers are made available for catalytic reactions at the material surface due to their spatial separation and long lifespan. When producing hydrogen, the separated electrons may efficiently make their way to the catalytic sites and convert protons ($H+$) to hydrogen (H_2). The efficiency of the hydrogen generation process is increased overall because of the efficient charge separation, which also allows the photocatalytic system to better use absorbed solar energy by converting more of it into chemical energy in the form of hydrogen [36].

6. Materials engineering for enhanced photocatalytic efficiency

Recent research has concentrated on van der Waals heterostructures, combining materials such as g-C_3N_4 and SiH, which have demonstrated potential in enhancing the photocatalytic characteristics required for effective hydrogen evolution. Van der Waals heterostructures are made up of two or more two-dimensional materials that are layered together using van der Waals forces rather than a strong chemical connection. Without the lattice mismatch problems that are typical of traditional heterojunctions, this special interaction makes it simple to tune electrical characteristics. The combination of different 2D materials, such as SiH (silicon hydride) and g-C_3N_4 (graphitic carbon nitride), can result in optimized band alignment [37]. This alignment is crucial for improving charge separation and transport, as it can be engineered to create a straddling band gap that facilitates efficient movement of electrons and holes to their respective reaction sites. These materials provide a high surface area and active sites crucial for the reaction. High surface area materials expose more catalytic sites to the reactants. In the case of photocatalytic water splitting, this means more available sites for water molecules to adsorb and subsequently be split into hydrogen and oxygen. Graphitic carbon nitride (g-C_3N_4) and silicon hydride (SiH) are two examples of 2D materials that can be combined to produce optimal band alignment [37]. It is possible to manipulate this alignment to produce a straddling band gap that efficiently transports electrons and holes to their respective reaction sites, which is essential for enhancing charge separation and transport. These substances offer a large surface area and active areas that are essential to the reaction. Materials with a large surface area reveal more catalytic sites to the reactants. More sites for water molecules to adsorb and then split into hydrogen and oxygen are thus accessible in the case of photocatalytic water splitting. Materials like g-C_3N_4 are known for their stable, polymeric structures that not only offer high surface areas but also possess suitable electronic properties for catalysis. The presence of nitrogen in g-C_3N_4, for instance, introduces p-type semiconductor characteristics that are beneficial for the adsorption and reduction processes involved in hydrogen production [38].

7. Photocatalytic water-splitting mechanism

The overall photocatalytic water-splitting process can be described by two half-reactions:

Oxidation Reaction (Oxygen Evolution Reaction, OER):

$$2H_2O \rightarrow O_2 + 4H^+ + 4e^- \tag{14}$$

Reduction Reaction (Hydrogen Evolution Reaction, HER):

$$4H^+ + 4e^- \rightarrow 2H_2 \tag{15}$$

For efficient water splitting, the photocatalyst must have a CB energy level more negative than the hydrogen reduction potential (0 V vs. NHE) and a VB energy level more positive than the oxygen evolution potential (+1.23 V vs. NHE). These reactions are driven by photoexcited electrons and holes generated upon light absorption by the photocatalyst.

8. Type-II, Z-scheme, and S-scheme heterojunctions

To optimize photocatalytic hydrogen production, several heterojunction designs, such as Type-II, Z-scheme, and S-scheme heterojunctions, are widely employed. Each scheme has unique mechanisms for enhancing charge separation and reducing electron-hole recombination, which are critical for improving the efficiency of photocatalytic processes (**Figure 3**) [39].

8.1 Type-II heterojunctions

Two semiconductors' valence band (VB) and conduction band (CB) align staggered in a Type-II heterojunction. While holes can go from one material's VB to another's VB, electrons can migrate from one semiconductor's CB to another's lower-energy CB, thanks to this band alignment. By reducing recombination, this spatial separation of electrons and holes promotes more effective charge use [40]. For photocatalytic hydrogen evolution, the g-C_3N_4/MoS_2 system is a well-studied example of a Type-II heterojunction [39]. MoS_2, a 2D material with appropriate band locations, and g-C_3N_4, a well-known photocatalyst with visible light activity, combine to generate a Type-II heterojunction that improves electron transport and raises the efficiency of the hydrogen evolution reaction

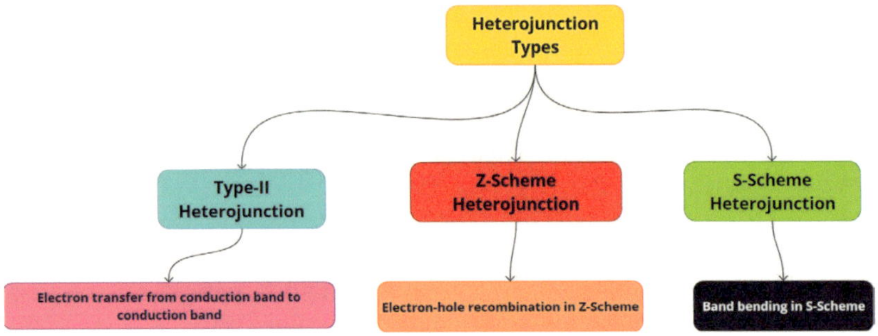

Figure 3.
Types of heterojunctions.

(HER). MoS_2's layered structure increases the surface area for catalytic reactions and adds more active sites. A ZnS/ZnSe Type-II heterojunction, which enhances hydrogen production because of its efficient hot photocarrier transfer mechanism, is demonstrated in another noteworthy study [41]. The effectiveness of photocatalytic hydrogen generation is greatly increased by the rapid electron transfer from ZnS to ZnSe made possible by the staggered band locations (**Figure 4**) [40].

8.2 Z-scheme heterojunctions

In the Z-scheme heterojunction, two distinct photocatalysts are connected to produce a stepped potential gradient, simulating the natural photosynthetic process. Unlike Type-II heterojunctions, Z-scheme systems allow for the recombination of photogenerated electrons and holes between two semiconductors. This process leaves behind highly energetic electrons and holes that are spatially separated, which improves both the hydrogen evolution reaction (HER) and oxygen evolution reaction (OER) [43, 44]. For example, the $ZnIn_2S_4$/NiCoP Z-scheme system demonstrated

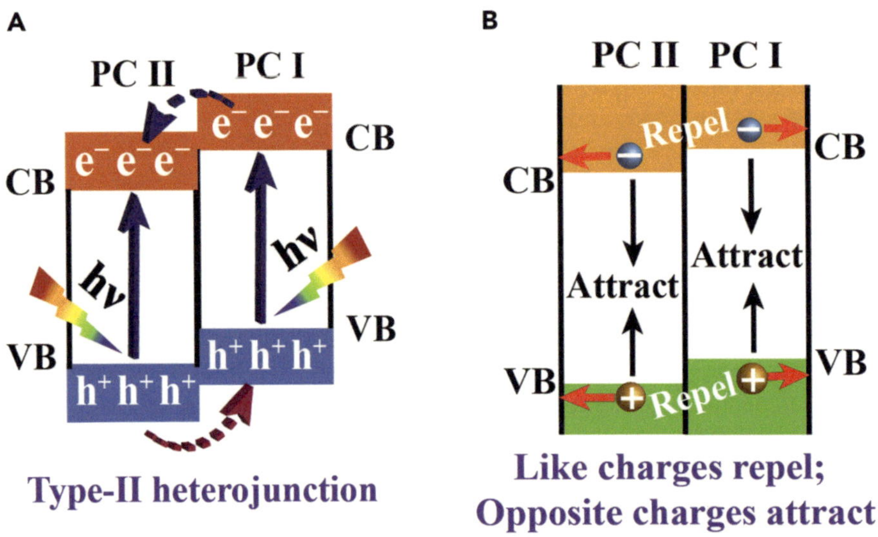

Figure 4.
Electron transfer mechanism in type-II heterojunctions and their limitations reproduced with permission [42].

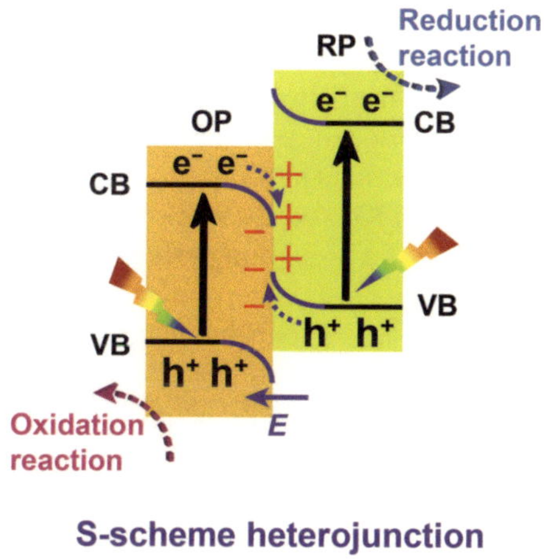

Figure 5.
S-scheme heterojunction schematic reproduced with permission [42].

enhanced visible-light-driven photocatalytic hydrogen production by improving electron-hole separation and reducing recombination rates. The hydrogen production rate in this Z-scheme heterojunction reached 5098.12 μmol.g^{-1}.h^{-1}, showcasing its superior performance compared to conventional photocatalysts. Another remarkable Z-scheme system is the g-C_3N_4/TiO_2 nanotube-based photocatalyst, where the interaction shifts from Type-II to Z-scheme heterojunctions as the ratio of g-C_3N_4 increases. This transition significantly enhances the photocatalytic hydrogen production rate due to improved charge separation [43].

8.3 S-scheme heterojunctions

The S-scheme heterojunction is an advanced version of the Z-scheme, designed to preserve strong redox potentials while improving electron transfer efficiency. An S-scheme system is more effective than conventional Type-II or Z-scheme heterojunctions because photogenerated electrons and holes are pushed to their appropriate catalytic sites without sacrificing the oxidation and reduction potentials. A recent study on ZnO/CdS S-scheme heterojunctions, for instance, showed increased photocatalytic hydrogen production as a result of better electron transfer efficiency and carrier separation. The hydrogen evolution rate is greatly increased by the handy conduit that the tubular structure of ZnO and CdS affords for photogenerated charges. Another study focuses on the Co9Se8/TiO_2 S-scheme heterojunction, which improved light absorption and charge separation made possible by the S-scheme design, resulting in a notable rise in H_2 evolution efficiency. The potential of S-scheme heterojunctions in large-scale hydrogen generation applications was demonstrated by this heterojunction, which performed better than the individual components (**Figure 5**) [45].

9. Work function compatibility, pH, oxidation, and reduction potentials, electronegativity

The photocatalyst and co-catalyst's energy levels must line up for effective photocatalytic water splitting. To attain maximum performance, factors including pH, electronegativity, oxidation and reduction potentials, and work function compatibility must be properly considered [45].

The energy needed to extract an electron from a material's surface to the vacuum level is known as the material's work function. The photocatalyst and the co-catalyst need to have compatible work functions in order to promote charge transfer. For instance, the conduction band of TiO_2 and the work function of platinum are well matched in a Pt-modified TiO_2 system, enabling effective electron transport to the Pt sites for hydrogen evolution [46].

The efficiency of charge transfer is also influenced by the difference in electronegativity between two materials. The difference in electronegativity between the two materials in a heterojunction system such as g-C_3N_4/TiO_2 promotes electron flow from TiO_2 to g-C_3N_4, increasing the rate of hydrogen generation [46].

The oxidation and reduction potentials of the water-splitting reaction are influenced by the electrolyte solution's pH. The efficiency of hydrogen synthesis is increased in an acidic environment because the reduction potential for hydrogen evolution is decreased. For instance, it has been demonstrated that raising the pH to a somewhat acidic level in S-scheme ZnO/CdS heterojunctions increases the efficiency of H2 production [47].

To ascertain whether the photocatalyst can propel the hydrogen evolution and oxygen evolution reactions, the oxidation and reduction potentials of water splitting are crucial. The photocatalyst's valence band needs to be more positive than the oxidation potential of H_2O/O_2 (~1.23 V vs. NHE), and its conduction band needs to be more negative than the reduction potential of H^+/H_2 (~0 V vs. NHE) [47]. The band alignment, for example, guarantees that the CB and VB positions are energetically favorable for both HER and OER in the V_2O_5/C_3N_5 Z-scheme heterojunction, leading to increased photocatalytic hydrogen generation. For effective charge separation and transfer in such systems, it is essential to match the Fermi energy levels and optimize the interface design [43].

A very successful method for improving charge separation and lowering recombination in photocatalytic H_2 production is the application of Type-II, Z-scheme, and S-scheme heterojunctions. The efficiency of solar-driven hydrogen production can be greatly increased by these heterojunction devices by adjusting parameters including pH, band alignments, electronegativity, and work function compatibility. Realizing large-scale hydrogen production using photocatalytic water splitting requires ongoing study into creating sophisticated heterojunction designs and comprehending their charge transfer mechanisms [45].

10. Properties of co-catalyst

10.1 Bandgap of co-catalyst

The band gap of the co-catalyst is an important parameter for HER. For effective H_2 production, the conduction band position of the co-catalyst should be aligned appropriately with the conduction band position of a photocatalyst. To facilitate the transfer of electrons, the conduction band energy level of the co-catalyst should be more negative than the conduction band of the photocatalyst but at the same time, it should be higher than hydrogen reduction potential to effectively carry the reduction reaction [48, 49]. The positions of the valance band and conduction band are also very crucial because a too-narrow band gap could cause the band position to drop below the potential needed for proton reduction which would make the photocatalyst inactive [50]. A catalyst with a narrow band gap can capture a wider range of solar light which improves light harvesting and increases photocatalytic activity [51]. For semiconductors, the ideal band gap is 1.23 eV to absorb visible light but practically it should be greater than 2.0 eV [52]. This makes it possible for the co-catalyst to produce enough charge carriers for photocatalytic reaction by absorbing more light from the solar spectrum. Band gap engineering of the covalent organic framework has shown that narrowing the band gap significantly improves photocatalytic activity. In the study of COFs, the COF-OH-3 exhibits the highest hydrogen production rate of 9.89 mmol. $g^{-1}.h^{-1}$. This is due to the optimal band gap of 2.28 eV, which efficiently utilizes solar energy and improves photocatalytic hydrogen production [53].

10.2 Surface-to-volume ratio of co-catalyst

The surface-to-volume ratio of the co-catalyst also affects the catalytic activity because it provides more active sites. Co-catalyst with more active sites reduces the activation energy for HER [54]. Improved catalyst-reactant contact made possible by larger surface area makes hydrogen production more effective. Studies have demonstrated that materials like silicon carbide (SiC) and other materials with large specific surface areas provide much better catalytic performance when accompanied by co-catalyst. A high surface-to-volume ratio allows better dispersion of catalyst nanoparticles which improves the catalyst-reactant contacts and raises the H_2 production rate [55]. Furthermore, the distinct nano-porous structures of these catalysts also enhance charge separation, durability, and stability during reaction [56]. Studies have shown that the porous structure of the co-catalyst maintains the catalytic activity for a long time and also prevents the agglomeration of catalyst particles which reduces the surface area and consequently the catalytic performance. For example, Shah et al. [57] reviewed about atomic scale structure of MoS_2 nanosheets and discussed the formation of its heterostructure with TiO_2, g-C_3N_4, and other semiconductors that boost H_2 production. Reducing the size of MoS_2 to quantum dots also resolves the issue of appropriate band alignment.

10.3 Work function of co-catalyst

Work function is the amount of energy required to remove an electron from the surface of a material. The work function of co-catalyst tells about the ease with which electrons can move from catalyst to 2D material heterostructure. Studies show that the modulation of the work function of the co-catalyst causes a redistribution of electrons spontaneously which creates a built-in electric field that promotes water adsorption and increases the charge transfer efficiency, which consequently increases the activity of HER [58]. Lower work function makes it easier for electrons to transfer from the conduction band of the photocatalyst to the co-catalyst. On the catalyst surface, H-intermediate adsorption energy optimization ensures efficient adsorption and desorption, which is critical for HER [58, 59]. Charge recombination also increases with a decrease in work function. Co-catalysts with high work function cause the trapping of photoelectrons, which is essential for photocatalytic activity. Beasley, Kumaran Gnanamani [60] studied different co-catalysts like Pt and Pd and compared their work function to that of TiO_2. It was found that the work function of Pt and Pd is higher compared to TiO_2, which results in excellent hydrogen production. The work functions of Ag, Ru, and Cu are either equal or less than TiO_2 due to which the excited electron moves back to TiO_2. Materials having higher work functions have better catalytic properties because they provide more active sites that facilitate the proton to reduce to hydrogen gas. Co-catalysts with suitable work functions also increase the stability of photocatalysts [61].

10.4 Strong interaction with 2D heterostructure

The robust interaction of co-catalyst with 2D heterostructure is a very important parameter to maintain the long-term stability and catalytic performance of 2D heterostructure for hydrogen production [57]. Co-catalysts strongly coupled with heterostructure efficiently transfer charges and reduce their recombination [62]. Strong interaction between co-catalyst and 2D heterostructure makes the system able to absorb a broader spectrum of solar radiations which leads to the generation of more charge carriers. The interaction between co-catalyst and 2D heterostructure produces more active sites [63]. For example, the introduction of a metal-organic framework increases the density of catalytic sites, which increases the catalytic performance by increasing the adsorption and activation of reactants [64]. DFT calculations have also shown that such type of interaction lowers the dissociation energy barrier of water molecules and increases the efficiency of hydrogen production [59]. Weak interaction results in the reduced generation and separation of photogenerated electron-hole pairs, inefficient absorption of sunlight, lower density of catalytic sites, high energy barrier, and also reduced stability and durability. Weak interaction also leads to poor charge transfer that causes the recombination of photogenerated electron-hole pairs before they participate in H_2 production [65].

10.5 Low overpotential

To produce H_2 gas, additional voltage is required to overcome inefficiencies like energy losses, slow reaction kinetics, etc. This additional voltage is known as overpotential. For efficient hydrogen production, the overpotential should be as low as possible. Various factors affect the overpotential such as limitations in mass transport and charge transfer, interfacial bonding, and kinetics barriers. Adding a co-catalyst in 2D heterostructure significantly reduces the overpotential. The addition of a co-catalyst produces more active sites and decreases the energy barrier, which consequently reduces the overpotential. It also improves the charge transfer efficiency between photocatalyst and co-catalyst, which is crucial because low charge transfer causes higher overpotential. For example, Yu et al. [66] incorporated MoS_2 into a 2D heterostructure, which facilitates charge transfer and increases electrical conductivity. Co-catalysts can stabilize reaction intermediates which ensures their performance over a long time, increasing reaction efficiency and lowering the need for extra voltage [67]. The electronic structure of the 2D heterostructure is also modified by aligning the energy levels in a more favorable position, as seen in MoS_2 heterostructure. This results in the optimization of the binding energy of the hydrogen atom, thereby reducing overpotential [68]. The optimal photodeposition method provides uniform distribution and robust interaction between co-catalyst and 2D heterostructure that increase the charge transfer efficiency, reduce the energy loss, and significantly reduce the overpotential, thereby increasing hydrogen production.

10.6 Chemical stability

The chemical stability of 2D heterostructure and co-catalyst is very crucial to maintaining the H_2 production efficiency for a long time. Chemical stability refers to the ability to resist any chemical change or degradation during an H_2 evolution reaction. When a co-catalyst is added, its efficiency is affected in many ways. The high amount of co-catalyst causes aggregation and blocks the active sites so the amount of co-catalyst should be optimized to increase stability without aggregation [57]. The active sites of heterostructure can be stabilized by co-catalyst, which ensures their activation throughout the reaction [69]. The stability can be enhanced by controlling the pH and temperature of the operating system. It can also be increased by using advanced techniques to form strong bonds and well-defined interfaces between co-catalyst and 2D material. Surface characteristics of 2D materials can also be modified by adding a co-catalyst, which increases the resistance of material from photo corrosion and from other degradation or leaching processes. Weak interfacial interaction between co-catalyst and photocatalyst causes leaching during chemical reactions, which increases the structural instability and loss of their performance during HER. Noble metals like Pt are highly stable and do not undergo corrosion for long-term use for hydrogen production. The co-catalyst and photocatalyst should be stable at high temperatures because they cause changes in the structural phase. For example, Xu et al. [70] prepared a ternary composite photocatalyst and found improvement in the production of hydrogen. This enhancement is due to a decrease in the recombination rate of photogenerated charge carriers and higher chemical stability and electrical conductivity. So, to enhance the long-term efficiency of hydrogen production, we need materials that have high chemical stability.

Author details

Durre Sameen[1], Saqib Ahmad[2] and Faryal Idrees[3*]

1 School of Physical Science, University of the Punjab, Lahore, Pakistan

2 Department of Physics, University of Sargodha, Sargodha, Pakistan

3 Department of Physics, University of the Punjab, Lahore, Pakistan

*Address all correspondence to: faryal.physics@pu.edu.pk

References

[1] Wang G, Chang J, Tang W, Xie W, Ang YS. 2D materials and heterostructures for photocatalytic water-splitting: A theoretical perspective. Journal of Physics D: Applied Physics. 2022;**55**(29):293002

[2] Novoselov KS, Geim AK, Morozov SV, Jiang DE, Zhang Y, Dubonos SV, et al. Electric field effect in atomically thin carbon films. Science. 2004;**306**(5696):666-669

[3] Shifa TA, Wang F, Liu Y, He J. Heterostructures based on 2D materials: A versatile platform for efficient catalysis. Advanced Materials. 2019;**31**(45):1804828

[4] Yoon JW, Kim JH, Kim C, Jang HW, Lee JH. MOF-based hybrids for solar fuel production. Advanced Energy Materials. 2021;**11**(27):2003052

[5] Zhang F, Dong Y, Jiang P, Wang G, Zhao N, Zhang H, et al. Highly dispersed and small-sized nickel (II) hydroxide Co-catalyst prepared by Photodeposition for hydrogen production. Chemistry–An Asian Journal. 2019;**14**(23):4193-4200

[6] Fareza AR, Nugroho FAA, Abdi FF, Fauzia V. Nanoscale metal oxides–2D materials heterostructures for photoelectrochemical water splitting—A review. Journal of Materials Chemistry A. 2022;**10**(16):8656-8686

[7] Mishra A, Sharma SK, Somvanshi D. 2D materials based heterostructure photocatalysts: Fundamentals and perspective. In: Materials for Electronic, Magnetic, and Spintronic Technologies: Characterization and Applications from Energy Storage to Disease Detection. Switzerland: Springer; 2024. pp. 21-39

[8] Pham PV, Bodepudi SC, Shehzad K, Liu Y, Xu Y, Yu B, et al. 2D heterostructures for ubiquitous electronics and optoelectronics: Principles, opportunities, and challenges. Chemical Reviews. 2022;**122**(6):6514-6613

[9] Jin L, Wang H, Cao R, Khan K, Tareen AK, Wageh S, et al. The rise of 2D materials/ferroelectrics for next generation photonics and optoelectronics devices. Apl Materials. 2022;**10**(6):060903-1-15

[10] Armor JN. The multiple roles for catalysis in the production of H2. Applied Catalysis A: General. 1999;**176**(2):159-176

[11] Wang Y, Ding Z, Arif N, Jiang W-C, Zeng Y-J. 2D material based heterostructures for solar light driven photocatalytic H 2 production. Materials Advances. 2022;**3**(8):3389-3417

[12] Zhang J, Hong H, Zhang J, Fu H, You P, Lischner J, et al. New pathway for hot electron relaxation in two-dimensional heterostructures. Nano Letters. 2018;**18**(9):6057-6063

[13] Idrees F, Butt FK, Hammouda SB. Progression in photocatalytic materials for efficient performance. MDPI. 2021;**11**(4):472

[14] Garcia JC, De Lima DB, Assali LV, Justo JF. Group IV graphene-and graphane-like nanosheets. The Journal of Physical Chemistry C. 2011;**115**(27):13242-13246

[15] Zhao A, Zhang L, Guo Y, Li H, Ruan S, Zeng Y-J. Emerging members of two-dimensional materials: Bismuth-based ternary compounds. 2D Materials. 2020;**8**(1):012004

[16] Lv X, Wei W, Sun Q, Li F, Huang B, Dai Y. Two-dimensional germanium monochalcogenides for photocatalytic water splitting with high carrier mobility. Applied Catalysis B: Environmental. 2017;**217**:275-284

[17] Pareek A, Dom R, Gupta J, Chandran J, Adepu V, Borse PH. Insights into renewable hydrogen energy: Recent advances and prospects. Materials Science for Energy Technologies. 2020;**3**:319-327

[18] Zou Z, Ye J, Sayama K, Arakawa H. Direct splitting of water under visible light irradiation with an oxide semiconductor photocatalyst. Nature. 2001;**414**(6864):625-627

[19] Li X, Bi W, Zhang L, Tao S, Chu W, Zhang Q, et al. Single-atom Pt as co-catalyst for enhanced photocatalytic H2 evolution. Advanced Materials. 2016;**28**(12):2427-2431

[20] Wang Y, Ou JZ, Balendhran S, Chrimes AF, Mortazavi M, Yao DD, et al. Electrochemical control of photoluminescence in two-dimensional MoS2 nanoflakes. ACS Nano. 2013;**7**(11):10083-10093

[21] Shahid W, Idrees F, Iqbal MA, Tariq MU, Shahid S, Choi JR. Ex situ synthesis and characterizations of MoS2/WO3 heterostructures for efficient photocatalytic degradation of RhB. Nanomaterials. 2022;**12**(17):2974

[22] Zhao A, Li H, Hu X, Wang C, Zhang H, Lu J, et al. Review of 2D group VA material-based heterostructures. Journal of Physics D: Applied Physics. 2020;**53**(29):293002

[23] Tahir M, Cao C, Mahmood N, Butt FK, Mahmood A, Idrees F, et al. Multifunctional g-C3N4 nanofibers: A template-free fabrication and enhanced optical, electrochemical, and photocatalyst properties. ACS Applied Materials and Interfaces. 2014;**6**(2):1258-1265

[24] Kumar A, Sharma P, Sharma G, Dhiman P, Shekh M, Sillanpää M, et al. Recent progress in advanced strategies to enhance the photocatalytic performance of metal molybdates for H2 production and CO2 reduction. Journal of Alloys and Compounds. 2023;**971**:172665

[25] Boretti A, Banik BK. Advances in hydrogen production from natural gas reforming. Advanced Energy and Sustainability Research. 2021;**2**(11):2100097

[26] Steinberg M. The Hy-C process (thermal decomposition of natural gas) potentially the lowest cost source of hydrogen with the least CO2 emission. Energy Conversion and Management. 1995;**36**(6-9):791-796

[27] Katebah M, Linke P. Analysis of hydrogen production costs in steam-methane reforming considering integration with electrolysis and CO2 capture. Cleaner Engineering and Technology. 2022;**10**:100552

[28] David M, Ocampo-Martínez C, Sánchez-Peña R. Advances in alkaline water electrolyzers: A review. Journal of Energy Storage. 2019;**23**:392-403

[29] Grigoriev S, Fateev V, Bessarabov D, Millet P. Current status, research trends, and challenges in water electrolysis science and technology. International Journal of Hydrogen Energy. 2020;**45**(49):26036-26058

[30] Aralekallu S, Lokesh KS, Singh V. Advanced bifunctional catalysts for energy production by electrolysis of earth-abundant water. Fuel. 2024;**357**:129753

[31] Mohan SV, Pandey A. Sustainable Hydrogen Production: An Introduction. Biohydrogen: Elsevier; 2019. pp. 1-23

[32] Alptekin FM, Celiktas MS. Review on catalytic biomass gasification for hydrogen production as a sustainable energy form and social, technological, economic, environmental, and political analysis of catalysts. ACS Omega. 2022;**7**(29):24918-24941

[33] Das BK, Hoque SN. Assessment of the potential of biomass gasification for electricity generation in Bangladesh. Journal of Renewable Energy. 2014;**2014**(1):429518

[34] Preethi V, Kanmani S. Photocatalytic hydrogen production. Materials Science in Semiconductor Processing. 2013;**16**(3):561-575

[35] Huang L, Gao R, Xiong L, Devaraji P, Chen W, Li X, et al. Two dimensional Ni 2 P/CdS photocatalyst for boosting hydrogen production under visible light irradiation. RSC Advances. 2021;**11**(20):12153-12161

[36] Nikolaidis P, Poullikkas A. A comparative overview of hydrogen production processes. Renewable and Sustainable Energy Reviews. 2017;**67**:597-611

[37] Ran J, Guo W, Wang H, Zhu B, Yu J, Qiao SZ. Metal-free 2D/2D phosphorene/g-C3N4 Van der Waals heterojunction for highly enhanced visible-light photocatalytic H2 production. Advanced Materials. 2018;**30**(25):1800128

[38] Zhang X, Gao D, Zhu B, Cheng B, Yu J, Yu H. Enhancing photocatalytic H2O2 production with Au co-catalysts through electronic structure modification. Nature Communications. 2024;**15**(1):3212

[39] Li J, Li M, Jin Z. 2D/3D ZIF-9/Mo15S19 S-scheme Heterojunction for productive Photocatalytic hydrogen evolution. Energy Technology. 2022;**10**(2):2100669

[40] Yao H, Jin G, Sui G, Li J, Guo D, Liang S, et al. ZIF-67-derived ZnIn2S4/NiCoP Z-scheme heterojunctions for enhanced visible-light-driven photocatalytic hydrogen production. Colloids and Surfaces A: Physicochemical and Engineering Aspects. 2022;**653**:129991

[41] Vempuluru NR, Kwon H, Parnapalle R, Urupalli B, Munnelli N, Lee Y, et al. ZnS/ZnSe heterojunction photocatalyst for augmented hydrogen production: Experimental and theoretical insights. International Journal of Hydrogen Energy. 2024;**51**:524-539

[42] Xu Q, Zhang L, Cheng B, Fan J, Yu J. S-scheme heterojunction photocatalyst. Chem. 2020;**6**(7):1543-1559

[43] Liu F, Zhang Q, Chen C, Zhang Z, Fang X. Theoretical design and experimental study a novel direct Z-scheme V2O5/C3N5 heterojunction for efficient photocatalytic hydrogen production. Solar Energy Materials and Solar Cells. 2023;**257**:112385

[44] Gao H, Cao R, Xu X, Zhang S, Yongshun H, Yang H, et al. Construction of dual defect mediated Z-scheme photocatalysts for enhanced photocatalytic hydrogen evolution. Applied Catalysis B: Environmental. 2019;**245**:399-409

[45] Lu H, Liu Y, Zhang S, Wan J, Wang X, Deng L, et al. Clustered tubular S-scheme ZnO/CdS heterojunctions for enhanced photocatalytic hydrogen production. Materials Science and Engineering: B. 2023;**289**:116282

[46] Wang Y, Fiaz M, Kim J, Carl N, Kim YK. Kinetic evidence for type-II Heterojunction and Z-scheme interactions in g-C3N4/TiO2 nanotube-based Photocatalysts in Photocatalytic hydrogen evolution. ACS Applied Energy Materials. 2023;**6**(10):5197-5206

[47] Wang X, Ge L, Wei K, Xing Z, Feng S. Fabrication of S-scheme Zn1-xCdxS/NiO heterojunction for efficient photocatalytic hydrogen evolution. International Journal of Hydrogen Energy. 2024;**55**:718-728

[48] Charles H, Pawar RC, Khan H, Lee CS. Photocatalyst engineering for water-based CO2 reduction under visible light irradiation to enhance CO selectivity: A review of recent advances. International Journal of Precision Engineering and Manufacturing-Green Technology. 2023;**10**(4):1061-1091

[49] Náfrádi M, Veréb G, Firak DS, Alapi T. Photocatalysis: Introduction, mechanism, and effective parameters. In: Green Photocatalytic Semiconductors: Recent Advances and Applications. Switzerland: Springer Nature; 2022. pp. 3-31

[50] Nasir JA, ur Rehman Z, Shah SNA, Khan A, Butler IS, Catlow CRA. Recent developments and perspectives in CdS-based photocatalysts for water splitting. Journal of Materials Chemistry A. 2020;**8**(40):20752-20780

[51] Ghosh S, Küçükkeçeci H, Paitandi RP, Weigelt V, Dippold V, Seki S, et al. Low band gap semiconducting covalent organic framework films with enhanced photocatalytic hydrogen evolution. Journal of Materials Chemistry A. 2024;**12**(1):247-255

[52] Bera S, Ghosh S. Bandgap engineering of heterostructures for visible light-driven water splitting. In: Green Photocatalytic Semiconductors: Recent Advances and Applications. Switzerland: Springer Nature; 2022. pp. 701-722

[53] Chen Y, Luo X, Zhang J, Hu L, Xu T, Li W, et al. Bandgap engineering of covalent organic frameworks for boosting photocatalytic hydrogen evolution from water. Journal of Materials Chemistry A. 2022;**10**(46):24620-24627

[54] Tong R, Ng KW, Wang X, Wang S, Wang X, Pan H. Two-dimensional materials as novel co-catalysts for efficient solar-driven hydrogen production. Journal of Materials Chemistry A. 2020;**8**(44):23202-23230

[55] Pinzón M, Sánchez P, de la Osa AR, Romero A, de Lucas-Consuegra A. Recent insights into low-surface-area catalysts for hydrogen production from ammonia. Energies. 2022;**15**(21):8143

[56] Li G, Tan Y, Lei Z, Yin F, He X. Preparation of high surface area carborundum-supported cobalt catalysts for hydrogen production by ammonia decomposition. Carbon Letters. 2023;**33**(3):899-908

[57] Shah SA, Khan I, Yuan A. MoS2 as a co-catalyst for photocatalytic hydrogen production: A mini review. Molecules. 2022;**27**(10):3289

[58] Li T-F, Li J, Zhang L-P, Ke J-W, Fan M-X, Zhang L-F, et al. Work-function-induced interfacial electron redistribution of MoO2/WO2 heterostructures for high-efficiency electrocatalytic hydrogen evolution reaction. Rare Metals. 2024;**43**(2):489-499

[59] Hu J, Liu X, Wang J, Jin J, Ouyang M, Fan M, et al. An asymmetrical Zr 2 CO/VSe 2 heterostructure as an efficient electrocatalyst for the hydrogen evolution

reaction. New Journal of Chemistry. 2024;**48**(31):13690-13699

[60] Beasley C, Kumaran Gnanamani M, Santillan-Jimenez E, Martinelli M, Shafer WD, Hopps SD, et al. Effect of metal work function on hydrogen production from photocatalytic water splitting with MTiO2 catalysts. ChemistrySelect. 2020;**5**(3):1013-1019

[61] Bai S, Yin W, Wang L, Li Z, Xiong Y. Surface and interface design in cocatalysts for photocatalytic water splitting and CO 2 reduction. RSC Advances. 2016;**6**(62):57446-57463

[62] Mehmood R, Ahmad Z, Hussain MB, Athar M, Akbar G, Ajmal Z, et al. 2D–2D heterostructure g-C3N4-based materials for photocatalytic H2 evolution: Progress and perspectives. Frontiers in Chemistry. 2022;**10**:1063288

[63] Feng W, Zhang W, Lin Q, Zhang H, Qiao J, Xia L, et al. Metal-support interactions of 2D carbon-based heterogeneous catalysts for hydrogen evolution reaction. Journal of Materials Chemistry A. 2024;**12**:18866-18878

[64] Ma Y, Fang H-X, Chen R, Chen Q, Liu S-J, Zhang K, et al. 2D-MOF/2D-MOF heterojunctions with strong hetero-interface interaction for enhanced photocatalytic hydrogen evolution. Rare Metals. 2023;**42**(12):3993-4004

[65] Zhang S, Bai B, Liu J, Zhang J. Atomically dispersed catalytic sites: A new frontier for cocatalyst/photocatalyst composites toward sustainable fuel and chemical production. Catalysts. 2021;**11**(10):1168

[66] Yu H, Zhang M, Cai Y, Zhuang Y, Wang L. The advanced progress of MoS2 and WS2 for multi-catalytic hydrogen evolution reaction systems. Catalysts. 2023;**13**(8):1148

[67] Li R, Xu H, Yang P, Wang D, Li Y, Xiao L, et al. Synergistic interfacial and doping engineering of heterostructured NiCo (OH) x-CoyW as an efficient alkaline hydrogen evolution electrocatalyst. Nano-Micro Letters. 2021;**13**(1):120

[68] Song Y, Xia K, Gong Y, Chen H, Li L, Yi J, et al. Controllable synthesized heterostructure photocatalyst Mo 2 C@ C/2D gC 3 N 4: Enhanced catalytic performance for hydrogen production. Dalton Transactions. 2018;**47**(41):14706-14712

[69] Zhang H-H, Ren Y-B, Yuan Z-L, Kang N-X, Mehdi S, Xing C-C, et al. Interfacial active sites on Co-Co2C@ carbon heterostructure for enhanced catalytic hydrogen generation. Rare Metals. 2023;**42**(6):1935-1945

[70] Xu J, Zhu J, Niu J, Chen M, Yue J. Efficient and stable photocatalytic hydrogen evolution activity of multi-heterojunction composite photocatalysts: CdS and NiS2 co-modified NaNbO3 nanocubes. Frontiers in Chemistry. 2020;**7**:880